Kilian Imfeld

Large-Scale High-Resolution Microelectrode Arrays Neurointerfaces

Kilian Imfeld

Large-Scale High-Resolution Microelectrode Arrays Neurointerfaces

System Design and Signal Processing

Südwestdeutscher Verlag für Hochschulschriften

Impressum/Imprint (nur für Deutschland/ only for Germany)
Bibliografische Information der Deutschen Nationalbibliothek: Die Deutsche Nationalbibliothek verzeichnet diese Publikation in der Deutschen Nationalbibliografie; detaillierte bibliografische Daten sind im Internet über http://dnb.d-nb.de abrufbar.
Alle in diesem Buch genannten Marken und Produktnamen unterliegen warenzeichen-, marken- oder patentrechtlichem Schutz bzw. sind Warenzeichen oder eingetragene Warenzeichen der jeweiligen Inhaber. Die Wiedergabe von Marken, Produktnamen, Gebrauchsnamen, Handelsnamen, Warenbezeichnungen u.s.w. in diesem Werk berechtigt auch ohne besondere Kennzeichnung nicht zu der Annahme, dass solche Namen im Sinne der Warenzeichen- und Markenschutzgesetzgebung als frei zu betrachten wären und daher von jedermann benutzt werden dürften.

Verlag: Südwestdeutscher Verlag für Hochschulschriften Aktiengesellschaft & Co. KG
Dudweiler Landstr. 99, 66123 Saarbrücken, Deutschland
Telefon +49 681 37 20 271-1, Telefax +49 681 37 20 271-0, Email: info@svh-verlag.de
Zugl.: Neuchâtel, Université de Neuchâtel, Diss., 2008

Herstellung in Deutschland:
Schaltungsdienst Lange o.H.G., Berlin
Books on Demand GmbH, Norderstedt
Reha GmbH, Saarbrücken
Amazon Distribution GmbH, Leipzig
ISBN: 978-3-8381-0473-7

Imprint (only for USA, GB)
Bibliographic information published by the Deutsche Nationalbibliothek: The Deutsche Nationalbibliothek lists this publication in the Deutsche Nationalbibliografie; detailed bibliographic data are available in the Internet at http://dnb.d-nb.de.
Any brand names and product names mentioned in this book are subject to trademark, brand or patent protection and are trademarks or registered trademarks of their respective holders. The use of brand names, product names, common names, trade names, product descriptions etc. even without a particular marking in this works is in no way to be construed to mean that such names may be regarded as unrestricted in respect of trademark and brand protection legislation and could thus be used by anyone.

Publisher:
Südwestdeutscher Verlag für Hochschulschriften Aktiengesellschaft & Co. KG
Dudweiler Landstr. 99, 66123 Saarbrücken, Germany
Phone +49 681 37 20 271-1, Fax +49 681 37 20 271-0, Email: info@svh-verlag.de

Copyright © 2009 by the author and Südwestdeutscher Verlag für Hochschulschriften Aktiengesellschaft & Co. KG and licensors
All rights reserved. Saarbrücken 2009

Printed in the U.S.A.
Printed in the U.K. by (see last page)
ISBN: 978-3-8381-0473-7

*to Jutta
and my parents*

Abstract

Microelectrode arrays (MEAs) - based neuronal interfaces allow monitoring and stimulating neuronal networks both *in vivo* and *in vitro*. *In vitro* methodology is widely used to study and model at intermediate complexity learning and memory processes within a large neuronal network. Commercially available MEA platforms for *in vitro* electrophysiology allow recording from about 100 electrodes. However, dissociated neuronal cultures can consist of up to 100'000 cells within a monitored area of a few square millimeters. Thus, substantial spatial subsampling compromises the observation of a large-scale network down to the cellular level. Therefore, higher resolutions are required and recently, researchers implemented MEA systems in CMOS-based technologies enabling the number of integrated electrodes to be largely increased. However, none of the current systems is compatible with both high-resolution and large-scale monitoring of dissociated neuronal cultures.

The objectives of this work are (i) the design of a complete large-scale and high-resolution MEA platform and (ii) the investigation and development of strategies to manage the large data streams with respect to efficient neurophysiological analysis. The developed system implements 4096 electrodes on an area of $7\,mm^2$. It was validated with cardiomyocytes as well as hippocampal and cortical neurons. Sampling frequencies up to $8\,kHz$ over the entire array lead to data flows in the order of 500 Mbit/s. Therefore, it is critical to provide a signal processing framework that enables fast on-line or real-time analysis of neurophysiological signals. The strategy, based on wavelet transforms, that was developed in this thesis enables a fast analysis framework that can converge towards a real-time implementation in hardware.

Wavelet transforms are well adapted to finite support signals such as neuronal spikes. Discrete wavelet transforms can be efficiently implemented using digital filter banks. This thesis demonstrates that all necessary functions for the analysis of neurophysiological signals, such as signal enhancement, spike detection and spike sorting, can be performed with fast wavelet transforms. Furthermore, a wavelet processor for a large number of electrode channels is designed in order to delineate the proposed methodology of hardware-based real-time processing of large-scale and high-resolution acquisitions of neurophysiological activity.

Activity monitoring using the large-scale high-resolution MEA platform motivates, for the first time, the use of signal processing techniques from the image/video field for the representation and characterization of neuronal networks. Increased spatial resolution can lead to observable spatial correlation of signals from adjacent electrodes, which, in turn, can be used for further signal enhancement. It turns out that spatial correlation and temporal information can not be independently accessed and that a more complete 3d-analysis is therefore necessary to benefit from the high spatio-temporal resolution. In addition, image-based representation of neurophysiological activity can potentially lead to new network characterisation methods. The multiresolution property of 2d-wavelet transforms is a possible approach for characterizing a neuronal culture at both network levels and at local/cellular levels. This concept is demonstrated by a multiresolution characterization of a network. In that sense, a center of gravity (CoG) is introduced in order to track the propagation of network bursts. Moreover, the CoG can be defined and tracked at different spatial resolutions allowing the characterization of networks at various levels of detail.

Contents

Abstract   i

1 Introduction   5

    1.1 Microelectrode Arrays - A System Overview 8

       1.1.1 Bio-Interface 9

       1.1.2 Signal Acquisition 15

       1.1.3 Signal Pre-Processing 15

       1.1.4 Spike Detection 17

       1.1.5 Network Analysis 18

    1.2 Problem Statement 19

    1.3 Objectives and Outline of the Thesis 20

2 High-Density Microelectrode Array Acquisition System   25

    2.1 System Specifications 26

       2.1.1 APS-MEA 26

       2.1.2 Data Acquisition Platform - Real-Time Signal Processing   28

2.2	System Implementation	29
	2.2.1 Architecture	29
	2.2.2 APS-MEA	32
	2.2.3 Acquisition Blocks	42
	2.2.4 Software	45
	2.2.5 Real-time Processing Units - Filters	47
2.3	Electrical Characterisation	50
	2.3.1 Single Electrode with Amplifier	50
	2.3.2 APS-MEA Acquisition System	52
2.4	Biological Measurements	54
	2.4.1 Reference Culture on Passive High-Density MEAs	56
	2.4.2 Dissociated Cardiomyocytes Cultures	58
	2.4.3 Dissociated Cortical Cultures	59
	2.4.4 Dissociated Hippocampal Cultures	61
2.5	Conclusion	65

3 Real-Time Signal Processing for High-Density MEA Systems 69

3.1	State-of-the-Art and Motivation	71
3.2	Denoising	74
3.3	Spike Detection	78

3.4	Spike Sorting	84
3.5	Conclusion	90

4 Image-Based Signal Processing   95

4.1	Signal Model	98
4.2	Redundancy of APS-MEA Signals	100
4.3	Spatio-Temporal Denoising	102
	4.3.1 Context	102
	4.3.2 Comparison of Multidimensional Denoising	103
	4.3.3 Discussion	107
4.4	Multiscale Analysis - Global vs. Local Activity	108
4.5	Conclusion	115

5 Hardware Implementation for Real-Time Signal Processing   121

5.1	Architecture	123
5.2	Validation	128
5.3	Application: Activity Monitoring on High-Density MEAs	130
5.4	Conclusion	133

6 Conclusion and Perspectives   137

A Mathematical Background - Wavelets   143

A.1 Introduction 143

    A.1.1 Wavelet Transform 143

A.2 Discrete Wavelet Transform 145

    A.2.1 Orthogonal Wavelet Transform 146

    A.2.2 Redundant Wavelet Transform 150

    A.2.3 Wavelet Packets 152

A.3 2d-Wavelet Transform 153

A.4 Summary 154

B Abbreviations   157

Bibliography   161

Publications Involving The Author   179

Acknowledgments   181

Chapter 1

Introduction

The human brain can probably be considered as one of nature's most complex systems. The studies aiming at uncovering the secrets of the human central nervous system by scientific methods started more than a century [1] ago. One important landmark in this field was the detailed model of the fundamental operation of a single neuron by Hodgkin and Huxley in 1952 [2], which led them to a Nobel Prize. Since then, using patch-clamp [3] and more recently also fluorescence imaging, many fundamental concepts that govern the single neuronal cell were discovered. Nowadays, complementary techniques are emerging (i.e. fluorescence resonance energy transfer) that will lead to an even better understanding of chemical processes occuring in the single neuron. However, single cell techniques allowing to simultaneously access a few neurons only, are not appropriate to investigate and unravel the function of the human brain that contains a highly complex network of 10^{12} interconnected cells. Therefore, in parallel to the cellular research, the brain at the organ level is now studied by structural and functional imaging. The former involves the use of, for example, computed tomography (CT) and diffuse optical imaging (DOI) while the latter exploits especially positron emission tomography (PET), functional magnetic resonace imaging (fMRI) and near infrared spectroscopic imaging (NIRS).

Notwithstanding the rapid advances in both, cellular and whole brain, technologies, the scientific community is still far from a thorough understanding of the brain. To deepen the knowledge of the mechanisms governing the brain, questions related to complex cognitive functions as well as learning and memorizing capabilities of neuronal systems have to be addressed. The study of neuronal networks consisting of thousands of interconnected cells, allowing to investigate the function and organization of large neuronal assemblies, is a valuable methodology to bridge the whole brain and single cell studies. One way to access such networks is by means of microelectrode arrays (MEAs). MEAs are a set of electrodes that measure the extracellular potential of neurons that are at or in a near proximity of an electrode. In comparison to intracellular measurement techniques such as patch-clamp, the MEA-based techniques are non-invasive, greatly simplify the experimental setup and importantly, allow simultaneous monitoring of many neurons over extended periods of time.

Over the last decade or so, the development of MEA systems focused on the implementation of several tens of electrodes on an area of a couple of square millimetres (Multichannel Systems[1], Panasonic, Ayanda-Biosystems[2], Tucker-Davis Technologies[3]). In recent years, the importance of increasing the number and the density of electrodes in the MEAs was acknowledged by neurophysiologists [4, 5] and the first high-density MEA concepts [6, 7] have been demonstrated. It has been shown that it was technologically possible to integrate thousands of electrodes on a surface of a few square millimetres. The importance of an efficient signal processing framework for the on-line or real-time analysis of the data at the architectural level was recognized but not yet addressed. In this respect, the development of signal and data analysis tools

---

[1] http://www.multichannelsystems.com
[2] http://www.ayanda-biosys.com
[3] http://www.tdt.com

currently lags behind the technological advances.

The use of microelectrodes for extracellular measurements of dissociated neuronal cells cultures is a promising methodology towards the understanding of learning and memory processes [8–11] and for the investigation of congenital behaviour in acute or cultured organotypic brain slices [12–14]. Moreover, MEAs are also extensively used in *in vivo* experiments [15]. By their means, it is possible to monitor large networks of interconnected neurons and thus to access the information processing capabilities of complex neuronal structures [16–18] at a mid- to long-term. Therefore, the MEAs bridge the gap in the understanding of the brain from a single neuron (i.e. by patch-clamp) to the high level behaviour of the whole brain (i.e. by functional imaging such as fMRI).

The ability to simultaneously access a large number of neurons permits to address fundamental studies such as the dynamic changes in distributed properties of networks and applied aspects such as effects of drugs and toxins. The study and evaluation of pharmacological active substances [19] or the implementation of biosensors [20] are just two examples of this development. It is well established that chemical modulators play an important role at the synaptic junctions [21] and are therefore key actors in the biological mechanisms that govern the information processing at the neuronal network level. Therefore, pharmacological tests using MEAs-based systems will certainly be one of the most significant applications. They will allow to enhance the information content and thus the efficiency of the drug screening assays aimed at selecting potential compounds for the treatment of neurodegenerative diseases. A subset of such systems consists of the screening of toxicological substances [22, 23].

A third field of interests for MEAs is biocybernetics, i.e. the study of communication, control and feedback applied to biological systems and prosthesis.

Neural interfaces [24–27] in the field of robotics and computing systems are subject to extensive study in order to access the highly robust operation, the massively parallel processing power and the auto-organizing structure of complex neuronal networks [28].

Finally, MEAs are also used for the validation of models of neuronal networks. Models are another methodology to reveal and understand the underlying physiological mechanisms [29–31]. Parameterization and validation of new models can typically be done by analyzing cells under specific and controlled conditions [32], which can be obtained and changed more readily in an *in vitro* environment than under *in vivo* conditions.

1.1 Microelectrode Arrays - A System Overview

Figure 1.1 depicts a system view of a generic MEA acquisition system. It contains all the important functional blocks from the biological cell up to the characterization of the network. A brief overview of the state-of-the-art of each block is given in the following subsections.

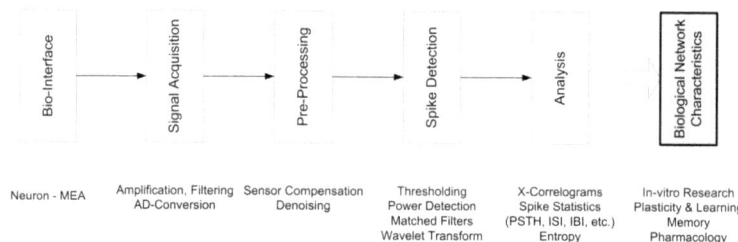

Figure 1.1 : A typical processing chain of MEA-based systems. A set of specific tasks (non exhaustive) for each block are listed. The first five blocks mainly concern engineering aspects whereas the last block refers to the relevance and interpretation of the biological experiments.

1.1.1 Bio-Interface

Neurons are electrically active cells. This stems from the fact that in the resting state there is a difference in concentration of ions in the cell with respect to that in the extracellular environment. For a given type of charge carrier, C, the equilibrium potential can be calculated by Nernst's equation [33]:

$$E_C = \frac{R \cdot T}{z \cdot F} \cdot \ln \frac{[C]_o}{[C]_i} \qquad (1.1)$$

where R is the thermodynamic gas constant $R = 8.314472 \frac{J}{K \cdot mol}$, T is the absolute temperature in Kelvin, z is the valence of the ion, F is the Faraday constant $F = 96485 \frac{C}{mol}$, and $[C]_o$ and $[C]_i$ are the molar concentrations of a particular ion outside (i.e. extracellular) and inside (i.e. intracellular) the cell, respectively.

The main ions to be considered are sodium (Na^+), potassium (K^+), calcium (Ca^{2+}) and chloride (Cl^-). Ion pumps (sodium-potassium and calcium pumps) ensure that there is a gradient between extracellular and intracellular concentrations i.e. the cell contains higher concentration of K^+ ions and lower concentrations of Cl^-, Na^+ and Ca^{2+} with respect to the extracellular medium. Taking into account the selective ion permeabilities through the cell membrane, these concentration gradients build up an electrical potential, i.e. the resting potential, of about - 65 mV (figure 1.2a).

Nernst's equation gives an equilibrium potential of - 80 mV for Potassium ions if the membrane is only permeable to K^+. However, this underestimates the value of the resting potential because it does not consider the small contributions of other selective ion gates (Na^+, Cl^-, Ca^{2+}) built into the cell membrane. The gates can be voltage- or ligand-sensitive (i.e. sensitive to ions or molecules),

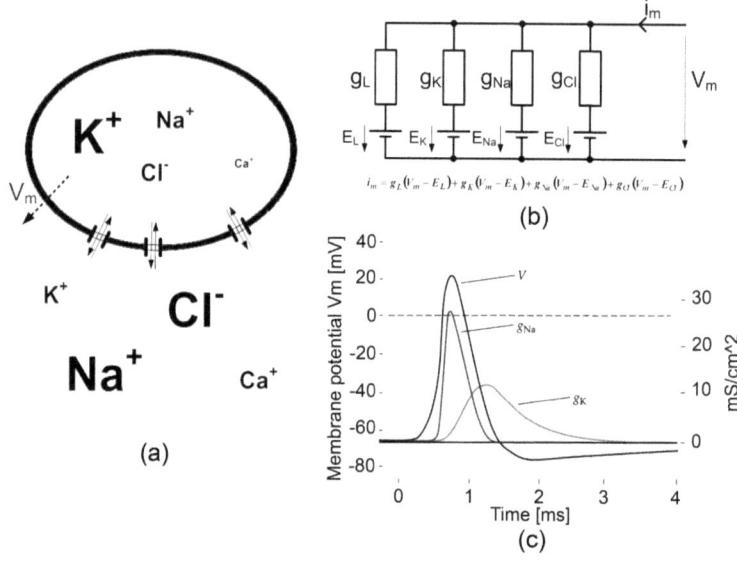

Figure 1.2 : The relative concentrations of the main ions inside and outside the cell. $[K^+]_o$=5mM, $[Na^+]_o$=150mM, $[Ca^{2+}]_o$=2mM, $[Cl^-]_o$=150mM, $[K^+]_i$=100mM, $[Na^+]_i$=15mM, $[Ca^{2+}]_i$=0.0002mM, $[Cl^-]_i$=13mM [34] (a) Hodgkin-Huxley-Model of a neuronal cell, E_K=-80mV, E_{Na}=62mV, E_{Cl}=-65mV are the membrane potentials if it is only permeable to K^+, to Na^+ or to Cl^- respectively. E_L corresponds to a lumped term for leakage currents and other contributions (i.e. E_{Ca}) (b). Typical shape of the membrane potential of an action potential (c) adapted from [35].

i.e. they change their state with respect to an applied voltage or with respect to the presence of a particular ion or molecule. It was Hodgkin and Huxley [2] in 1952 who found out the existence of selective ion gates (channels) in the cell membrane. Furthermore, they discovered that the conductance of these gates can change with respect to the voltage across the membrane. In the resting state these ion gates are closed. A simplified description of their model is shown in figure 1.2b. The leftmost branch in the equivalent electrical circuit (g_L, E_L) lumps together the small contributions of Ca^{2+} and other leakage currents.

An extracellular event (i.e. a stimulus) will change the cell membrane potential. When the voltage across the membrane crosses a given threshold, the

Na^+ channels open, setting off the influx of Na^+ ions (into the cell). The membrane is depolarized, i.e. the voltage goes from -60 mV up to 30 mV. This is then followed by the repolarization which is the result of the opening of the voltage-sensitive K^{2+} channels and the efflux of K^{2+} ions. Figure 1.2c shows a typical shape of a neuronal membrane potential also called a spike. After each fired action potential there follows an enforced silent period (i.e. refractory period) during which another impulse can not be fired. The action potential is a transitory event lasting about 1 ms with the refractory period of about 2 to 3 ms [34].

Neurons are interconnected through synapses. An action potential of a neuron propagates down its axon. The axon terminates in many small branches whose stubs are the synapses. The arriving electrical potential makes vesicles in the synapse move to the cell membrane. These vesicles release neurotransmitters that cross the synaptic gap (figure 1.3). The postsynaptic neuron has ligand-sensitive gates in its membrane and the resulting flow of ions generates a local variation in potential at that location. The branches where a postsynaptic neuron connects to its presynaptic neurons are called dendrites. It is the local variation of the membrane voltage from various synapses that is integrated and that can lead to the generation of a new action potential in the soma of the postsynaptic neuron. Thereafter, the action potential propagates again down the axon of the currently active neuron until it reaches the synapses where the same process starts again.

The human brain contains approximately 10^{12} neurons and each neuron can make thousands of connections to other neurons through synapses. Nowadays, it is not yet possible to study networks of that size and therefore smaller networks, such as they can be found in both *in vitro* cultures and *in vivo*, enable studies towards the understanding of the physiology of large neuronal network systems at an intermediate complexity [36, 37]. In that sense, microelectrode arrays (MEA) are one way to interact with networks containing thousands of cells [38, 39].

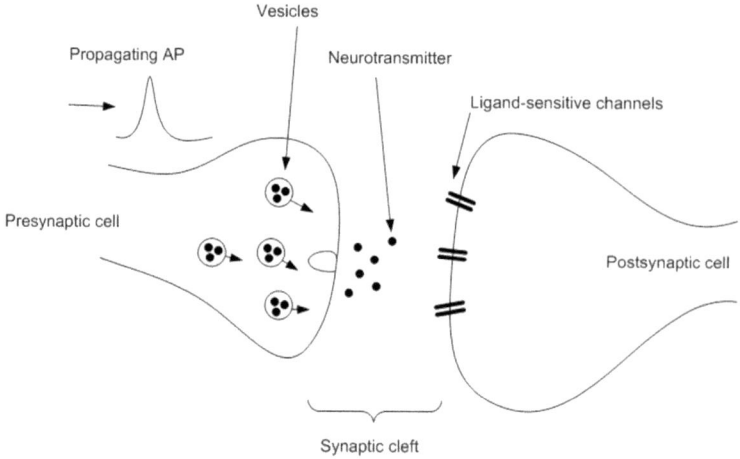

Figure 1.3 : The vesicles in the presynaptic cell move towards the cell membrane when an AP is arriving. They release neurotransmitters into the synaptic cleft. The neurotransmitters bind onto ligand-sensitive gates at the postsynaptic cell. As a consequence, the gates open and ions start crossing the membrane of the postsynaptic cell to contribute to the generation of an AP in this cell.

MEAs consist in a set of electrodes that are in contact with the cells and the culture medium. They measure the extracellular potential changes that arise at the interface cell-electrode. The resulting electrochemical interface is very complex and many studies have been carried out to understand and model its behaviour [33, 40–44]. A simplified, but widely used model of the interface is shown in figure 1.4.

Extracellular and intracellular measurements are intrinsically different and consequently, the shape and the amplitude of the recorded signals are distinct (figure 1.5). The major disadvantage of extracellular measurements with respect to intracellular ones is the three orders in magnitude decrease of the signal amplitude. A typical signal amplitude of an extracellular potential is in the order

1.1 Microelectrode Arrays - A System Overview

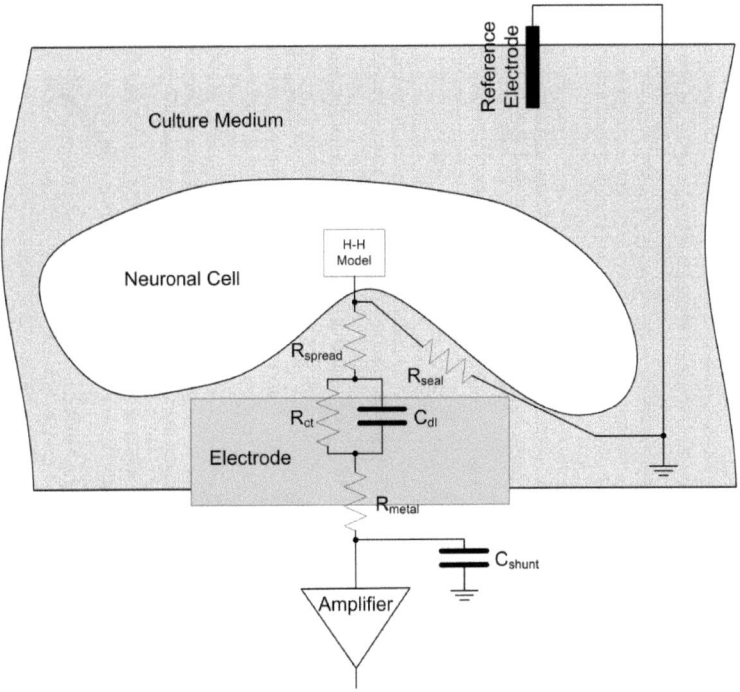

Figure 1.4 : The electrical model of the electrochemical interfaces electrode-electrolyte-cell [40]. R_{metal} is the resistance of the electrode, C_{shunt} consists of all the parasitic capacitors at the input of the amplifier. R_{ct} models the charge transfer at the metal-electrolyte interface and C_{dl} is the double layer capacitance at the interface [33]. R_{spread} is the resistor of the electrolyte between the cell and the electrode, R_{seal} is the sealing resistor due to the adhesion of the cell to the substrate. Models and measured values for R_{ct}, C_{dl}, R_{spread} and R_{seal} can be found in literature [42, 44–47].

of a few tens to a few hundreds of microvolts, depending on the cell type and the quality of cell/electrode interface [48]. The electronics has therefore to be very sensitive to detect these small signals.

Already in the seventies Thomas [49] used MEAs to interface with electrogenic cells. In the early eighties, Pine [39] and Gross [50] started investigating the behaviour of neuronal cultures systematically with MEAs and since then many research activities based their methods on the use of MEAs. Commercially available systems (Multichannelsystems, Ayanda, Panasonic) are based on mi-

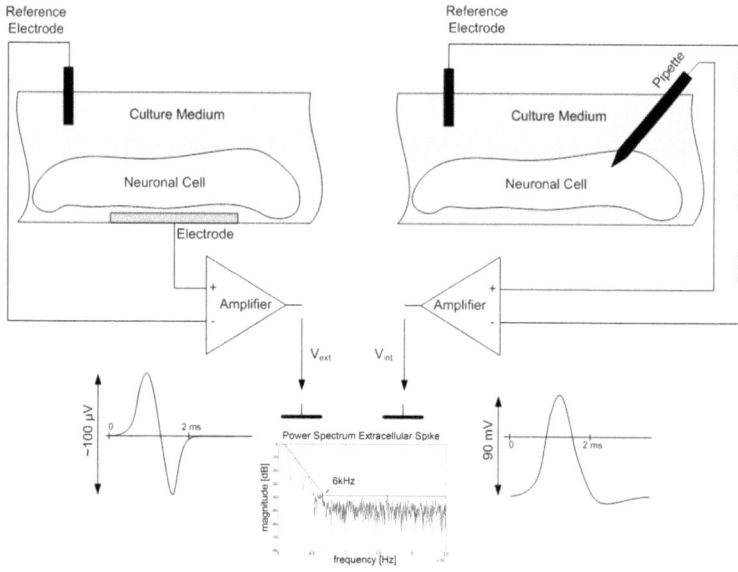

Figure 1.5 : Extracellular measurement (left), intracellular measurement (right). The difference in shape and amplitude of the recorded signal can be seen. Moreover, the extracellular technique is non-invasive which can enable long-term recordings.

cromachined substrate which integrate a number of passive electrodes with metallic connections [47]. The number of electrodes in such an array is limited to 60 - 128, in some prototypes the arrays contain 512 electrodes [51]. However, this passive approach is very impractical for even higher number of electrodes since the signals have to be routed to external amplifiers. Therefore, multiplexing of electrodes and integrating signal processing blocks on the same substrate as the electrodes using commercially available CMOS technologies [52–55] have been proposed. This allowed implementing up to 256 electrodes with amplifiers and filters on a single chip [56–60]. An alternative approach was introduced by Berdondini and Eversmann [7,61]. Berdondini designed an MEA consisting of 4096 metallic electrodes, however the system was limited to a concurrent readout of only a few electrodes. Eversmann implemented an array with 16384 electrodes using field effect transistor (FET) [62] electrodes. However, the achieved noise level is still too high to measure small extracellular spikes of mammal neuronal cells and his system was validated on cultured brain slices [63].

1.1.2 Signal Acquisition

Figure 1.6a shows a conventional MEA acquisition system and figure 1.6b illustrates the concept of multiplexing in a CMOS-based MEA system. Analog-to-digital conversion can either take place on-chip or externally by discrete electronic components. The same holds for the digital interface. Hafizovic [64] includes a field programmable gate array (FPGA) to perform a threshold-based standard spike detection [65] before sending the data to a host computer. Oweiss [66] proposes a hardware-implemented wavelet transform for data compression of 32 *in vivo* electrodes. Guillory [67] proposes a PC-based system for real-time spike detection and storage of 100 channels and Folgers [68] implemented a 128 channel wavelet transform on a dedicated digital signal processor (DSP).

1.1.3 Signal Pre-Processing

Signal processing issues not only arise for the analysis of biological data but particularly at the transducer level. All types of transducers experience non-idealities due to inherent systematic errors, random errors and external interferences. The effects of many of these factors can be attenuated or even eliminated by appropriate methods. General techniques such as linear time-invariant filtering [69] are easy to use and they can significantly improve the sensor performance when they are included in the system. They perform optimally when the spectral components of the errors do not overlap with spectral components of the useful signal. If this is not the case, statistic modelling of

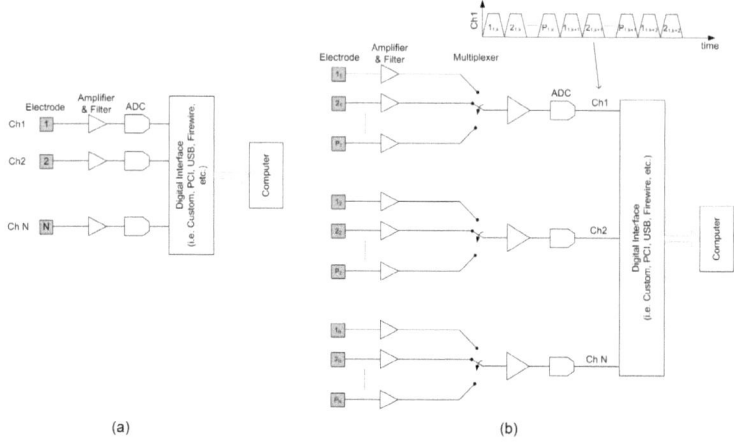

Figure 1.6 : Conventional MEA acquisition system with a signal chain for each electrode (a). Multiplexed MEA system as frequently used in solid-state circuits [61, 64]. One channel contains several electrodes that are interleaved in time using a switch (b). Analog-to-digital converters (ADCs) are used in both systems to digitize the analog waveforms. Thereafter, the samples are sent to a host computer for processing.

errors can still lead to substantial functional improvement [70]. In general, any signal processing method that improves the signal quality at the output of a system is referred to as signal enhancement or restoration. In most commercial MEA systems, the pre-processing is usually limited to rejecting out-of-band noise components. The system's bandwidth is tailored to the bandwidth of the signals to be measured. Therefore, band-pass filters are generally an integrated part very early in the signal chain of conventional MEA systems.

The use of microelectrode arrays enables multivariate statistical techniques to improve the signal quality of each individual channel. The key idea is to use redundant information from the same physical processes measured on different electrodes in order to improve the overall signal-to-noise ratio (SNR) of all channels [71–73]. principal component analysis (PCA) or independent component analysis (ICA) [74–77] are two such methods. They have already been widely used in other array processing fields [78], such as communication, speech and image processing [77]. These statistical methods have also been introduced in neurocomputing many years ago [79,80], mainly for *in vivo* systems and experiments. Oweiss recently proposed an *in vivo* multi electrode signal processing framework to enhance the quality of the recorded signals and to improve the discrimination of neuronal spikes from the electrode array [81].

1.1.4 Spike Detection

The fundamental information element in neuroelectrophysiology is the occurrence of a spike from a neuron. Researchers are mainly interested in the timing of a spike whereas the waveform shape is not considered as a significant parameter. The most important task in signal processing related to MEA systems is therefore the detection of spikes from a series of measurements from a set of electrodes. Thereafter, the spike events from all measured neurons are fed to statistical algorithms that aim to find any embedded deterministic behaviour. For many years, numerous mathematical methods to analyze spike trains from multielectrodes exist already. Since spike trains are event-related signals, discrete signal theory can not be applied as such and a different mathematical formalism, also referred to as point processes [82, 83], was developed to track event-based signals mathematically.

Extracellular neuronal spike amplitudes are in the range of several tens of microvolts to several hundreds of microvolts [40], depending on the type of neuronal cells. Cortical, hippocampal and spinal cell cultures can give typical signal amplitudes from $\approx 10\,\mu V_{pp}$ to $300\,\mu V_{pp}$. Therefore, MEA acquisition systems need to be highly sensitive, i.e. they must provide a very low system noise floor in order to detect extracellular biological signals. Many different methods have been explored [84] to detect neuronal spikes under low SNR (i.e. $0\,\text{dB} - 6\,\text{dB}$)

conditions. Currently, one of the most promising approaches for the detection of low SNR signals was achieved by using wavelet theory [85–89].

With conventional low-density MEAs multiple shapes can often be observed from one single electrode [5]. It is commonly agreed that the different signal shapes are the contribution of different neurons. Therefore, the number of observed neurons can be increased by the fact that it is possible to monitor more than one cell at one electrode. In order to discriminate between the neurons a set of characteristics (i.e. features) has to be extracted from each waveform. This can be achieved by statistical methods [90], such as PCA [91], or by wavelet-based methods [87, 92].

1.1.5 Network Analysis

Biological characterization of neuronal networks is mostly assessed by simple statistical measures derived from the spike trains (i.e. point processes). Practically, various histograms from one or multiple spike trains are constructed to estimate measures, such as inter spike interval probability density functions or cross-correlograms of multiple spike trains, respectively [82]. The bin width that is chosen to build the histograms is critical in order to prevent artefact relationships between the spike trains under investigation due to the fact that successive bins after a firing can be correlated because of the refractory period of this neuron. Usually, bin widths of 0.02 ms - 0.5 ms [93] are used for the computation of the histograms. From the probability density functions and cross-correlograms one can estimate whether there is a statistical dependence between the underlying processes (i.e. direct synaptic connection between neurons or indirect functional dependence) or whether the processes tend to be independent. Some other characteristics of the neuronal network can be revealed, such as synchronicity, pacemaker cells and functional connectivity between pairs of neuron [8], however more complex multivariate patterns remain undiscovered [5, 94]. The same statistical measurements can be applied between the stimulus trains and the corresponding response trains. In that case, we refer to post-stimulus-time-histogram (PSTH) which indicates the statistical dependence of a network to a stimulus and its (evoked) responses.

An alternative approach to assess the functional relationship between the observed neurons of a network is by means of information theoretic concepts. Entropy and mutual information measures are estimated from the spike events and serve to describe the functional connectivity of the network [13, 37, 95, 96]. These concepts can be extended to multivariate signals and can therefore quantify more complex underlying functional connectivity than with simple pairwise statistics as discussed above.

1.2 Problem Statement

Dissociated large-scale neuron cultures consist of 10'000-100'000 neurons on an active area of typically 6 mm^2 [97]. In order to observe detailed activity propagation at a local level it is important to have high-density recording sites, i.e. at least electrodes at distances in the order of cell dimensions. Therefore, the importance of high-density MEAs is evident since a few years [4]. However, next to monitoring the activity at a local level, it is also important to monitor the entire network for a significant statistical representation of the culture [5]. Both these constraints, high-resolution and large-scale, necessarily lead to MEAs with thousands of concurrent recording electrodes in order to bridge the gap between local activity patterns and global network characteristics [98]. Berdondini introduced the concept of high-density MEAs in 2001 [7], however the system was not laid out to concurrently record from all electrodes. Recently, other high-density systems in both conventional passive technology (see section 1.1.1) and CMOS-based technology [52, 56, 59, 61, 63, 99] were implemented. Eversmann achieves both high-density and large-scale recording, however the noise performance of the circuit ($250\,\mu V_{rms}$) does not allow the recording of spikes and bursts of dissociated cortical cultures from vertebrates and the acquisition system is not extended to continuous acquisition and real-time capability. Low SNR is an inherent characteristics of high-density large-scale CMOS MEAs since the noise performance decreases with an increase in electrode density, i.e. less silicon area per electrode hampers a low noise performance in the amplifiers [100].

Two major challenges of large-scale high-density MEA systems are therefore i) handling of the large amount of data that is produced during the recording and ii) the detection and analysis of signals in a low SNR environment. Classical signal analysis (i.e. spike detection, spike sorting, functional connectivity by pairwise cross-correlograms) becomes a difficult and tedious task, since data streams in the order of GBit/s are expected. On-line or real-time analysis of the electrophysiological data is even impossible with currently available acquisition systems (see section 1.1.2) and analysis tools [5]. New innovative solutions on both hardware system and signal processing levels are therefore required to face the challenges of new generations of large-scale high-density MEA systems.

1.3 Objectives and Outline of the Thesis

The main objective of this work is to implement and explore a large-scale high-density microelectrode array system that can continuously record spike and bursts from dissociated neuronal cultures. To achieve this, a new architecture for the MEA acquisition system will be introduced by taking into account the need for real-time signal analysis. The entire system will be presented as a new unified framework that enables both hardware- and software-based signal processing (figure 1.7). Then, real-time pre-processing of the electrophysiological data and the use of hardware to perform real-time spike detection and sorting will be addressed. The developed framework can be extended to use methods stemming from the image processing field. This will be illustrated by an alternative representation, processing, interpreting and analyzing of the physiological data based on concepts that are used in the field of image/video processing.

Chapter 2 of this thesis discusses the concept and design of large-scale high-density MEA acquisition systems. The architecture and the implementation of the system will be shown. Important aspects for real-time capabilities will be pointed out. Electrical and biological validation will then be described and acquisitions from neuronal cell cultures will be demonstrated.

Chapter 3 presents the concepts related to signal processing adapted to high-density MEA systems. In particular, it will be shown that signal denoising, spike detection and sorting can be done in a wavelet-based framework that can be extended to the real-time processing of neurophysiological signals.

Chapter 4 extends the multichannel approach of the signal processing framework to an image-based representation of the signals from high-density MEAs. This chapter will show that the high-density feature of our system leads to local spatial signal correlation. By using methods from the image processing field, this spatial correlation can be used for further enhancing the signals and improving the discrimination of spikes. Moreover, the multiresolution property of the wavelet decomposition can lead to new representations and characterizations of the activity of dissociated neuronal cultures.

The concept of the wavelet-based framework for real-time analysis leads to the implementation of a real-time wavelet transform kernel in hardware on an FPGA. Architecture, design and validation will be shown in chapter 5.

Chapter 6 concludes the thesis and gives a few perspectives that arise from the image/video interpretation of neurophysiological signals. A set of techniques from classical image processing opens the door for further research in the field of neurophysiology.

1.3 Objectives and Outline of the Thesis

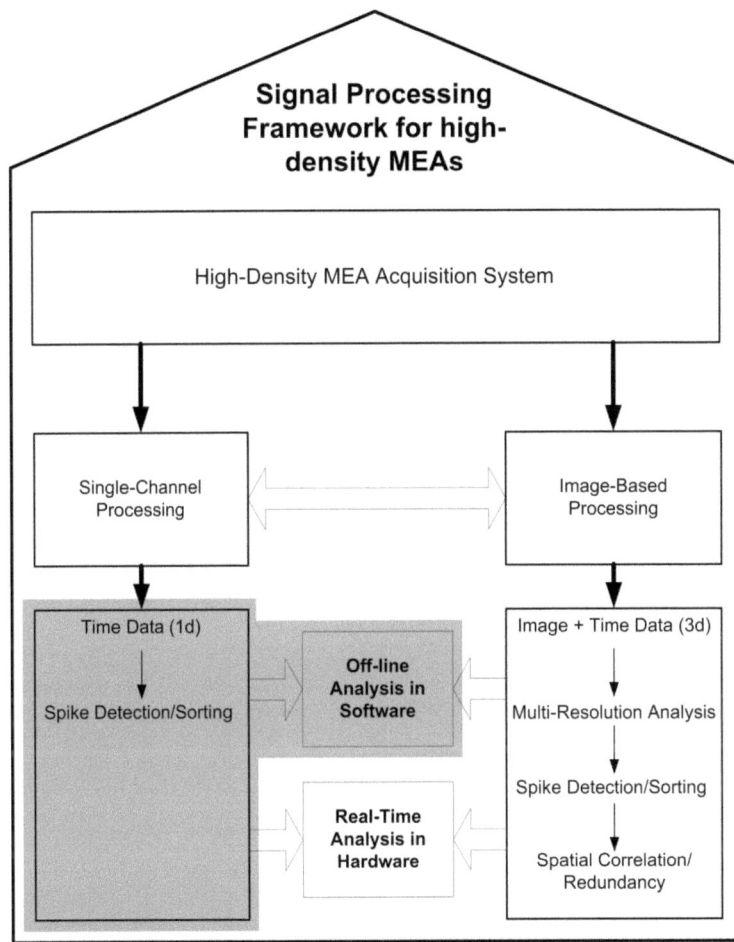

Figure 1.7 : The proposed framework includes the high-density MEA acquisition system and the signal processing concepts based on both single-channel and image-based representations. Single-channel processing shall be based on the analysis of multiple one-dimensional signals f(t) and image-based processing shall be performed with three-dimensional data f(x,y,t) (i.e. space and time). The gray shaded area corresponds to the conventional approach of multielectrode / microelectrode array systems.

Chapter 2

High-Density Microelectrode Array Acquisition System

Commercially available MEA systems feature sampling rates up to 50 kHz, noise levels lower than 3-5 μV_{rms}, signal bandwidths of 3 kHz and integrate typically 60-120 microelectrodes of 10-30 μm in diameter with pitches in the order of hundreds of micrometers. Assuming average cardiomyocyte cell lengths in the order of 60 μm, observations of subcellular propagation effects can not be achieved with current electrode densities. Furthermore, with typical neuron soma dimension in vertebrates of a few micrometers and typical neuronal networks of 10'000-100'000 neurons, the limited number of electrodes and their rather large pitch results in a substantial spatial undersampling of the overall network activity. The development of MEA systems featuring a higher spatial resolution while preserving adequate noise and sampling performances is thus a prerequisite for improving both the subcellular resolution capabilities in cardiomyocyte cultures and the local statistical activity representation of large organized neuronal populations.

Specifically addressing the high spatial resolution for *in vitro* devices, two main approaches of densely integrated electrode arrays [7] and field-effect transistor

(FET) arrays [61] were reported. In the latter FET-based approach, spatial resolutions down to 7.8 μm were achieved on arrays of 128 x 128 elements. As yet, the temporal resolution of 2 kHz full frame rate and up to 8 kHz on smaller areas and a noise level of 250 μV_{rms} do not measure up to that of conventional MEA systems [63]. A third approach of high-density MEAs was proposed by Frey [101]. This system can record from 126 configurable electrodes of a matrix of 11016 electrodes, however due to the limited number of recording sites it can not simultaneously acquire from a large-scale network at a high resolution. A hybrid approach implementing an external ASIC amplification chip for 512 electrodes that was recently reported [51] is not an effective solution for arrays of several thousands of electrodes.

Another aspect in developing high spatial resolution MEAs concerns the data acquisition system which in its current standard format (channel-by-channel addressing) is inadequate. The implementation of the real-time signal processing in hardware becomes thus essential for the handling of the large amount of data resulting from thousands of parallel recording sites.

In this chapter, we report on the implementation and validation of a complete electrophysiological recording platform, consisting of an APS-MEA device and an acquisition system with real-time signal and data processing resources based on architectural ideas from the image/video acquisition field.

2.1 System Specifications

2.1.1 APS-MEA

The APS-MEA shall comprises 64 x 64 electrodes arranged as an array of pixel elements. The electrode size is 21 μm x 21 μm since this dimension turned out

2.1 System Specifications

to be a good trade-off between the spatial resolution requirement and the microelectrode performances [99]. The distance between electrodes is 21 µm, so that an area of 42 µm x 42 µm is available underneath each electrode for basic in-pixel analog processing as enabled by commercial CMOS processes. The overall active area of the electrode array is 2.67 mm x 2.67 mm, which results in an electrode density of 576 electrodes/mm^2. It is worth noting that typical cell densities in cortical cultures range from 500 - 2500 cells/mm^2 resulting in a near one-to-one match between number of neurons and available electrodes [102].

In order to maximally avoid the occurrence of any unwanted electrochemical reaction at the interface electrolyte - CMOS-chip, a biasing of the electrolyte at 0 V with respect to the system ground is required.

Since signals from different types of electrogenic cells, and therefore of different amplitudes, shall be acquired, the dynamic range of the system needs to be very high. On the one hand, typical spike amplitudes of dissociated cortical neurons from rats are in the order of 100 μV_{pp}. On the other hand, signals up to several millivolts are obtained from cardiomyocytes cultures. Therefore, a programmable gain has to be implemented for adapting the acquisition system to this wide range of amplitudes. The sensitivity of the system will be given by the minimum noise floor that can be achieved. The noise performance of CMOS circuits is directly related to the available area and to their power consumption. Due to the high resolution requirement, both available area and maximum power consumption per electrode are very limited. Given the power supply at 3.3 V and the minimum and maximum signal amplitudes, i.e. respectively 100 μV_{pp} and 2 mV_{pp}, the minimum and maximum gain of the system can be determined.

2.1.2 Data Acquisition Platform - Real-Time Signal Processing

A two orders of magnitude increase of the number of recording sites with respect to the number of passive sites available in conventional MEAs entails severe constraints towards the mining and analysis of the recorded data. It is therefore imperative to integrate sufficient resources in the system in order to implement signal processing tasks. Non-idealities (i.e. drifts and offsets) of the APS-MEA compromise efficient signal analysis tasks at later stages of signal processing. Hence, it is necessary to perform signal enhancement, such as filtering, prior to applying further analysis. Filtering of 4096 channels in real-time needs an outstanding speed performance in the involved processing blocks and only hardware-implemented functions (i.e. FPGA) can meet these stringent requirements.

Both high spatial resolution at a large scale and high spatio-temporal resolution at a local level are required for the acquisition system. The bi-modal imaging capability, i.e. full-frame and zoomed modes, is a key feature for enhancing the correlation and the investigation of signal propagation at the macro- and micro-scale of the cellular network circuitry. The sampling rates for these two modes have to target a sufficient temporal resolution for performing efficient spike detection at the full frame level and even provide higher frequencies for micro-scale propagation analysis. Upon evaluation of both aspects on custom designed passive MEAs with high-density electrodes integrated in small areas [103], a 12-bit Analog-to-Digital (AD) conversion with a global sampling rate of 10 kSamples/s in the full frame mode and a minimum sampling rate of 20 kSamples/s in the zooming mode were defined. A bandwidth requirement of about 500 Mbit/s results from the specifications of the ADCs and the required spatio-temporal resolution.

Table 2.1 : Summary of basic specifications of the acquisition platform

Number of Electrodes	64 x 64
Electrode Size	21 μm x 21 μm
Electrode Element Area	42 μm x 42 μm
Supply Voltage	3.3 V
Gain Range	52 dB - 76 dB
Overall Power Consumption of Chip	132 mW
Input Referred Noise	15 μV_{RMS}
Maximum Zoom Mode Electrode Number	32 x 32
Full Frame Sampling Rate	10 kHz
Minimum Zoom Mode Sampling Rate	20 kHz
ADC Resolution	12 bit
Signal Frequency Range	1 Hz - 5 kHz

2.2 System Implementation

2.2.1 Architecture

Multiplexing has been used for more than a century in communications systems to transfer voices or videos of multiple users. This technique has also been widely applied in the optical sensing field, where light sensitive devices based on the concept of active pixel sensors (APS) [104–107] are a valuable alternative to the traditional charge-coupled devices (CCD). The same technique can also be successfully implemented to switch many on-chip integrated electrodes on one or several output channels and therefore it can decrease routing complexity. By replacing the photodiode of a pixel by an electrode, an APS-based architecture can be designed to implement a high-density MEA [7, 99]. Furthermore, applying the concepts and methods originally developed for image/video processing, an alternative approach to handle the data from a high-density MEA system can be found. Instead of acquiring a parallel set of one-dimensional signals along the amplitude vs. time quantities, the main idea behind the system organization, and that is reflected by the hardware implementation and signal handling, consists in representing the electrophysiological data as time sequences of images.

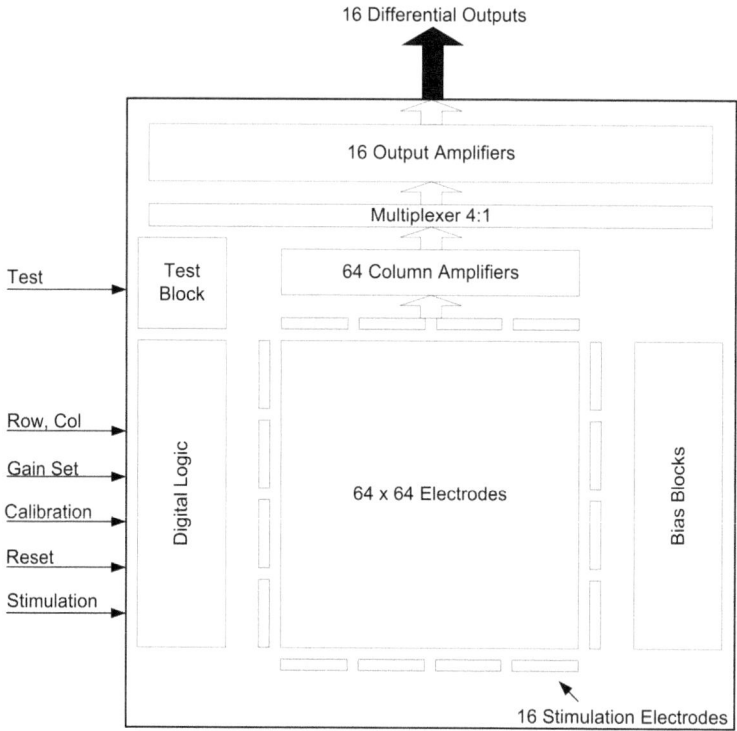

Figure 2.1 : APS-MEA Block Schematics

The front-end of the system consists thus of the APS-MEA featuring an array of electrodes with corresponding amplification, addressing and multiplexing functionalities (figure 2.1). The location of each electrode is precisely defined in terms of row and column indexes and all the electrodes are subsequently multi-

2.2 System Implementation

plexed on one or several parallel output channels. It is the on-chip multiplexing that enables the read-out of a large number of channels without being limited by the electrode interconnections. Therefore, multiplexing leads to the implementation of large-scale and dense electrode arrays for high spatial resolution.

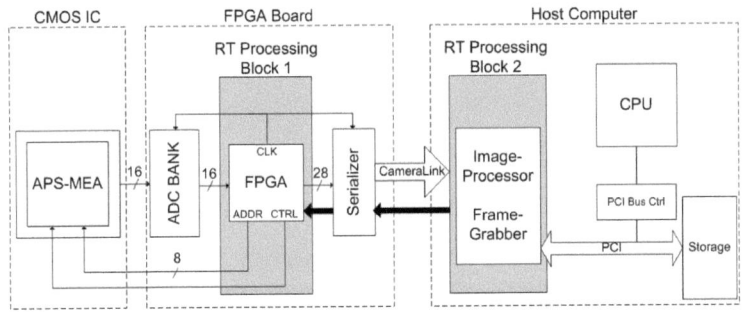

Figure 2.2 : Block diagram of the acquisition architecture. The high-speed communication interfaces are depicted with white arrows, whereas the low-speed ones going from the host computer to the FPGA are indicated by black arrows. The real-time (RT) blocks indicate the real-time processing units to be used for data pre-processing and analysis tasks. It has to be noted that the CPU does not actively participate in the high-speed data transfer.

The multiplexing of numerous electrodes on one single output entails stringent requirements with respect to the bandwidth along the critical paths of the integrated circuit. In order to relax these requirements, the signals of 4096 electrodes are multiplexed on 16 parallel analog output channels (i.e. 256 electrodes per channel). The complete architecture of the acquisition system is shown in figure 2.2. The 16 analog output channels from the APS-MEA are externally sampled by a bank of ADCs. The different outputs are sent to a FPGA through a serial interface internally provided by the ADCs. The FPGA is both the control and timing device of the APS-MEA and of the bank of ADCs, since the addressing of the electrodes has to be synchronized with the AD conversion. The interface between computer and FPGA is bidirectional, so that control data can be sent from the user interface, which allows for a high flexibility in the configuration of the hardware and of the experimental settings.

Pre-processing such as filtering is performed on the same FPGA. The data are then multiplexed to a high-speed interface and sent to the PCI frame grabber containing a dedicated image processor (RISC) that can be used for spike detection. Thereafter, the processed data is buffered on the frame grabber and subsequently sent to a hard drive by PCI bus. It is important to note that the general purpose CPU of the host computer only controls the setup configuration and the visualization of data. However, it is not included in the real-time chain. This architecture, stemming from high-speed camera systems, enables efficient pre-processing (e.g. real-time filtering) of the data at the hardware-level (i.e. in the FPGA). The pre-processing can be followed by real-time implementations of higher-level functions, such as wavelet-based signal compression [108], spike detection [85,89], spike sorting [92] and classification on either the FPGA or the image processor (RISC) of the frame grabber. A wavelet-based pre-processing implementation will be demonstrated in chapter 5.

2.2.2 APS-MEA

For high-sensitivity systems that have to reliably record signals in the range of $100\,\mu V_{pp}$, it is imperative to maximally reduce the noise contributions. Pre-amplification of the signals at the very beginning of the acquisition chain is essential in order to be more robust to the noise contributions of subsequent stages and to on- and off-chip noise coupling over the interconnection paths. Therefore, the APS-MEA consists of an array of pixel elements that comprise each an electrode, a pre-amplifier and a set of electronic components to bias and address the pixel. Furthermore, eight different bias current settings for the electrode element can be selected in order to test the circuit under different bias conditions. A photograph of the APS-MEA fabricated in a Austriamicrosystems (AMS) $0.35\,\mu m$ CMOS, 4 metal layers technology can be seen on figure 2.3.

2.2 System Implementation

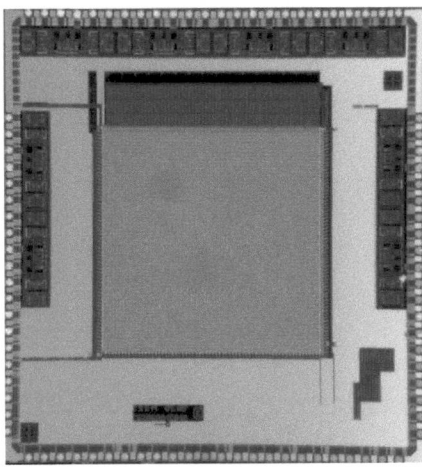

Figure 2.3 : Photograph of the APS-MEA V1 chip. The 4096 electrodes array is in the centre. The structures on the right, on the left and on the top border are the output drivers. The column amplifiers are visible at the top of the array.

Noise

Amplifying the signal as much and as early as possible in the chain to rise it above the total system noise floor on the one hand and limiting the maximum noise bandwidth on the other hand are the only two ways of increasing the SNR. In a conventional CMOS implementation of such an electrode amplifier, the 0V-bias of the electrolyte is another challenge when using an asymmetric power supply (0 - 3.3 V). We can either include an AC coupling at the electrode or perform a direct DC coupling. AC coupling requires an additional area consuming capacitor (e.g. metal-insulator-metal). Amplification can then be performed at the electrode (figure 2.4a). If the amplifier is DC coupled, a level-shifter is first needed to raise the voltage to a level that allows enough voltage headroom for the following amplification and bias stages (figure 2.4b).

A simplified model of the electrode including its noise contribution is shown in figure 2.5 (derived from the model shown in figure 1.4).

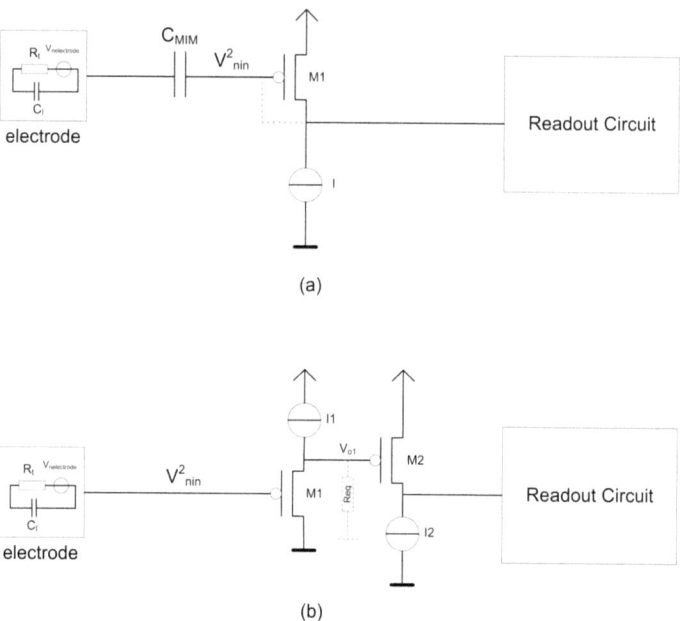

Figure 2.4 : AC coupled electrode with immediate amplification in a common-source amplifier. The DC input voltage is removed by a metal-insulator-metal capacitor (C_{MIM}). The dashed line symbolizes the circuitry that is necessary to bias the amplifier (a). DC coupled electrode with PMOS level-shifter and common-source amplifier in a second stage (b).

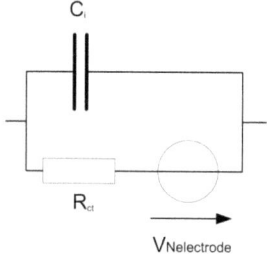

Figure 2.5 : Simplified electrode model including noise. C_i models the interface (i.e. metal-electrolyte) capacitor and R_{ct} models the charge transfer at the interface. $V_{Nelectrode}$ represents the noise voltage of the electrode.

2.2 System Implementation

The major noise sources of the electrode and the input amplifier are:

- Thermal noise of the electrode
- Thermal noise of the CMOS transistor(s)
- Flicker noise of the CMOS transistor(s)

The thermal noise of the electrode can be modelled as $V_{Nelectrode} = \sqrt{4 k R_N \Delta f}$ where R_N corresponds to the electrode's equivalent noise resistance and $\Delta f = 1/(R_{ct} \cdot C_i)$. For a platinum electrode of 10 µm of diameter, typical values of R_{ct}=430 kΩ and C_i=7.85 pF are obtained [47].

The thermal noise power density of MOS transistors can be modelled as follows:

$$\overline{I}_{nth}^{2} = \frac{2}{3} \cdot 4\,kT \cdot g_m \qquad (2.1)$$

with $\overline{I}_{nth}$ the noise current per $\sqrt{Hz}$, $k = 1.3806504 \cdot 10^{-23} \frac{J}{K}$ the Boltzmann constant and g_m the small signal transconductance of the MOS transistor [100].

Flicker noise power density of MOS transistors is inversely proportional to the product of their lengths L and widths W:

$$\overline{V}_{in\frac{1}{f}}^{2} = \frac{K}{C_{ox} \cdot W \cdot L} \cdot \frac{KF \cdot I^{AF}}{f} \qquad (2.2)$$

$$\overline{I}_{n,\frac{1}{f}}^{2} = \frac{K}{C_{ox} \cdot W \cdot L} \cdot \frac{1}{f} \cdot g_m^2 \qquad (2.3)$$

and thus directly related to the area of the pixel element. $\overline{V}_{in\frac{1}{f}}^{2}$ corresponds to the input-referred noise voltage, $\overline{I}_{n,\frac{1}{f}}^{2}$ to the flicker noise current in the transistor and I to the bias current of the transistor. f is the frequency and KF and AF are two dimensionless parameters that are fit to the frequency characteristics of the CMOS process. PMOS transistors are inherently less noisy than NMOS [100]. A quantitative characterisation of the flicker noise contribution of a single PMOS transistor in the AMS 0.35 µm CMOS process with respect to two different widths W is shown in figure 2.6.

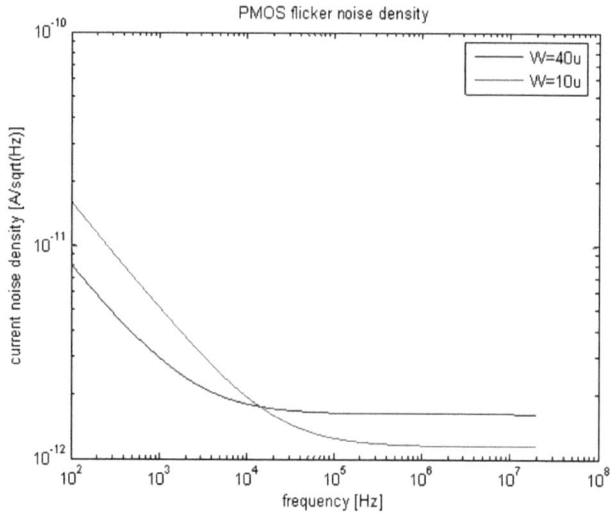

Figure 2.6 : Flicker noise for PMOS with L=0.5 µm and I=10 µA of the AMS 0.35 µm CMOS process ($KF = 1.191 \cdot 10^{-26}$ and $AF = 1.461$)

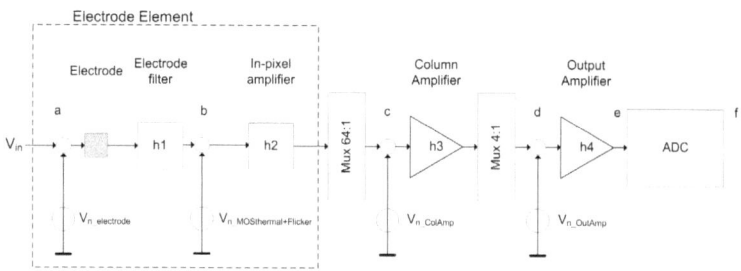

Figure 2.7 : System Transfer Functions

The multiplexing of many electrodes on one channel increases proportionally the bandwidth requirements of the channel signal path. The bandwidth of the multiplexed channels has to be at least the bandwidth of one individual pixel

2.2 System Implementation

multiplied by the number of multiplexed electrodes. Therefore, sampling of each channel will substantially alias high-frequency noise into the signal band and considerably degrade the overall noise performance, since the sampling frequency is only a fraction of the channel bandwidth [61]. Consequently, the total effective noise increases. A block diagram that identifies the noise-relevant transfer functions of the architecture is depicted in figure 2.7. It refers to the APS-MEA architecture of figure 2.1 and the system architecture of figure 2.2. A qualitative representation of the signal and noise spectra at each node of figure 2.7 is shown in figure 2.8. The power densities (PD) are normalized by the total gain at each node. The cut-off frequency of the electrode transfer function h1 is supposed to be much higher than the bandwidth of the electrode element (h2).

A qualitative analysis of both structures of figure 2.4 reveals a difference in noise performance. Neglecting the noise contribution of the readout-circuit and of the current source I, for the AC-coupled architecture the main noise source other than the electrode is the MOS transistor M1. By taking into account the folding mechanism as explained in figure 2.8, the input-referred square noise voltage for the AC coupled architecture (figure 2.4a) after sampling can be expressed as:

$$\overline{v}^2_{AC,Nin_{sampled}}(f) = \sum_{k=\lfloor\frac{-BW}{f_s}\rfloor}^{\lceil\frac{BW}{f_s}\rceil} \left(\frac{4kTR_N}{\left|1+j\cdot\frac{2\pi\cdot(f-k\cdot f_s)}{\omega_c}\right|^2} + \frac{\overline{I}^2_{M1,N_{th}(f-k\cdot f_s)} + \overline{I}^2_{M1,\frac{1}{f}(f-k\cdot f_s)}}{g^2_{m_{M1}}} \right) \qquad (2.4)$$

where $\omega_c^{-1} = R_t C_t$ corresponds to the bandwidth of the electrode, BW is the system noise bandwidth and f_s is the sampling rate.

Assuming that the gain of the first stage is approximatively 1, the DC-coupled

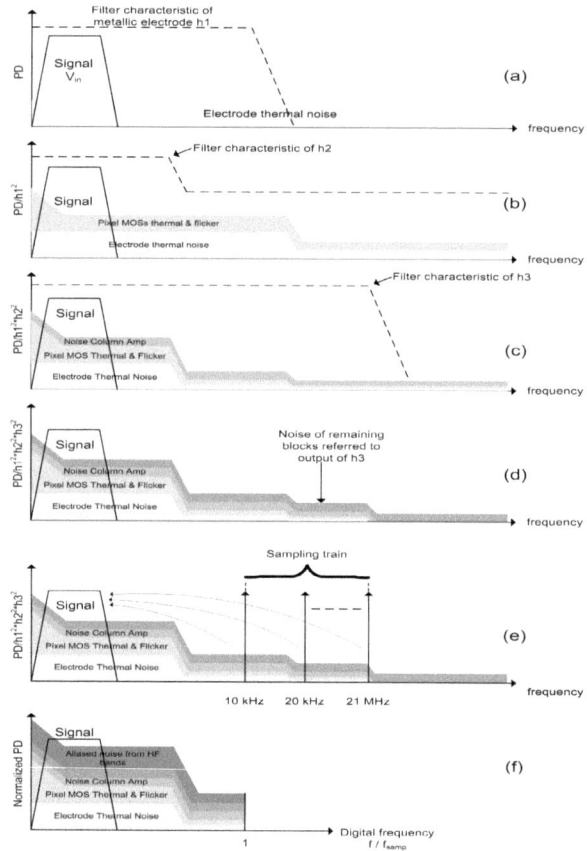

Figure 2.8 : Signal and noise power densities (PD) at the input of the electrode are shown (a). A typical electrode bandwidth is 50 kHz. Both filtering characteristic and noise densities at the input of the electrode amplifier (h2, cut-off frequency $f_c = 10kHz$)are added in (b). The same at the input of the column amplifiers (c) and the output amplifiers (d). The sampling of the spectrum is represented by a spike train up to the maximum channel bandwidth at distances that correspond to the sampling frequency (e). The digital spectrum is displayed after conversion showing the contribution of the down-folding due to the high-frequency (HF) bands around each Dirac function of the spike train (f). The noise contribution of each block is modelled with respect to its input, the PDs are normalized at each stage for illustrative purposes and the sampling train intervals are not in scale.

2.2 System Implementation

architecture (figure 2.4b) leads to:

$$\overline{v}^2_{DC,Nin_{sampled}}(f) = \sum_{k=\lfloor\frac{-BW}{f_s}\rfloor}^{\lceil\frac{BW}{f_s}\rceil} \left(\underbrace{\frac{4kTR_N}{\left|1+j\cdot\frac{2\pi\cdot(f-k\cdot f_s)}{\omega_c}\right|^2}}_{Electrode\ Contribution} + \frac{\overline{I}^2_{M2,N_{th}(f-k\cdot f_s)} + \overline{I}^2_{M2,\frac{1}{f}(f-k\cdot f_s)}}{g^2_{m_{M2}}} + R^2_{eq}\cdot\left(\overline{I}^2_{M1,N_{th}(f-k\cdot f_s)} + \overline{I}^2_{M1,\frac{1}{f}(f-k\cdot f_s)}\right) \right) \quad (2.5)$$

The last term in equation 2.5 adds a substantial noise contribution to the total noise, since R_{eq} (i.e. the impedance at the node V_{o1} of figure 2.4b) is usually much greater than $1/gm_{M2}$, and assuming that the noise currents of both transistors are about the same.

One way to reduce noise aliasing, i.e. reduce the number k of Dirac functions within the system bandwidth in equations 2.4 and 2.5, is to limit the maximum bandwidth per channel, or equivalently, the number of pixels multiplexed on one channel. This consists in dividing the entire array into a number of smaller zones each one having its own output channel. The current circuit implementation with 16 output channels was found to be a good trade-off between bandwidth limitation and increasing external routing complexity.

Electrode Element

The electrode element implements an amplification stage (operational transconductance amplifier, OTA) with a gain of 40 dB within the electrode element in order to mask the noise of subsequent stages. Moreover, a calibration stage was included in the pixel to limit the effects of DC offsets, thermal drifts and leakage currents arising from the dynamic electrochemical equilibrium at the electrode-electrolyte interface, which could saturate the following analog processing blocks. Due to limited pixel area, a capacitive AC coupling in the signal path was not feasible. Therefore, a DC coupled architecture was chosen to solve this offset problem (figure 2.4b). Consequently, a DC offset compensation circuitry within the electrode element [55] is required and is implemented in our device by an auto-zeroing structure, based on a sampled feedback (SW_{AZ1} in figure 2.9). During operation, a calibration sequence (SW_{AZ1} and SW_{AZ2} closed) lasting about $4\,\mu s$ is used for resetting the input to a defined potential of 0 V. This calibration biases the feedback accordingly. At the end of the calibration sequence the feedback path is opened and the sampling capacitor C_{AZ} holds the corresponding voltage value. The OTA then works in open-loop mode. However, due to leakage currents in the capacitor, ambient light, temperature changes and variations at the interfacial potential, the sampled voltage drifts and the in-pixel circuit has to be regularly recalibrated. The lower cut-off frequency of the signal band can be set according to the frequency of calibration. Under standard operation a typical calibration period of 1 to 2 s should be sufficient.

The amplification of the signal chain is completed by a programmable column amplifier, which allows to set a total gain of 52 dB, 64 dB, 70 dB or 76 dB, respectively, and to adapt the gain to signals from different cell cultures. Besides the nominal bias currents and in order to have additional tuning capabilities, different programmable bias currents for the in-pixel amplifier, read-out amplifier and output amplifier were also implemented. In-pixel low-pass filtering limits the total noise power at the output of the system (see figure 2.8b) [59], however again the tight area constraints due to the high spatial resolution requirement

2.2 System Implementation

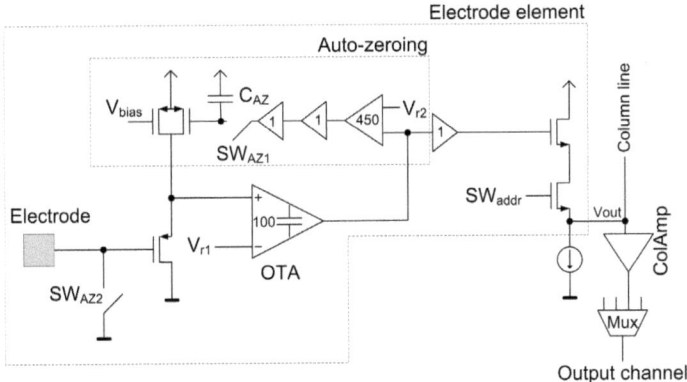

Figure 2.9 : Electrode element showing the signal flow to the column amplifier. There is one amplifier per column. Four column amplifiers are then multiplexed to one output channel. One channel itself includes another gain stage (12 dB) which serves as driving stage for the output.

do not allow efficient filter implementations for the required frequency bands. Nevertheless, some inherent low-pass filter characteristic in the pixel can be obtained by connecting a Miller capacitor between the outputs of the two input transistors of the OTA (figure 2.10). The noise of the electrode element will be determined by the electrode, the input transistor (level-shifter) and the OTA (see equation 2.5). The simulated noise voltage within the bandwidth of the electrode amounts to $15\,\mu V_{RMS}$.

A common-mode noise insensitive differential signal chain was implemented from the column amplifier on, however at the pixel level a single-ended signal path was preferred to a differential one. A differential topology requires a larger area and increases routing complexity, however it remains a promising alternative for a next-generation device in a smaller CMOS technology.

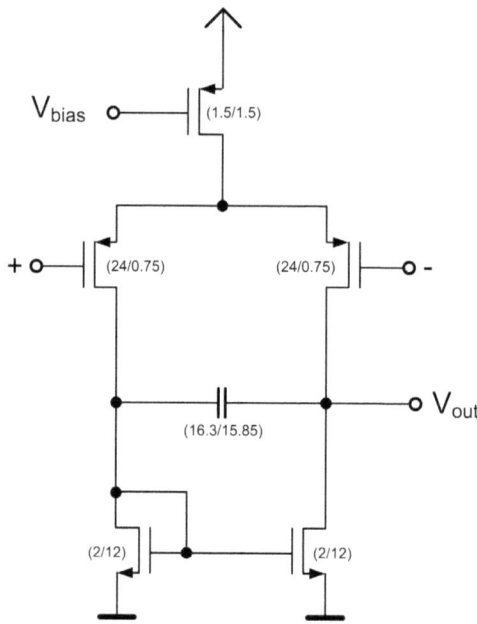

Figure 2.10 : In-pixel OTA. The floating capacitor was implemented using a PMOS transistor. The values between parentheses correspond to the width/length (in μm) of the transistors [schematics provided by S. Neukom, CSEM].

2.2.3 Acquisition Blocks

On-chip ADCs were avoided for the two following reasons: (i) design speed-up of the APS-MEA chip and minimization of design risks and (ii) rejection of any additional on-chip digital noise sources. Thus, the 16 APS-MEA differential output channels have to be digitized before further data processing can be performed. This is done on an intermediate system block that precedes the FPGA but is physically part of the FPGA circuit board (figure 2.11). Thus, a bank of 16 ADCs feeds the digital signals to the FPGA. The FPGA is an intrinsic component of the system architecture and enables hardware implementation of electrophysiological data analysis (see figure 2.2). This is essential for the real-time handling of a large number of electrodes. The expected data stream

2.2 System Implementation

at the output of the 12-bit ADCs amounts to 500 Mbit/s for the read-out at 10 kFrames/s of the entire array. Important but less hardware consuming tasks implemented on the FPGA are the control of the APS-MEA, such as configuration and calibration, as well as the control and timing of the addressing and of the ADC sampling. The FPGA is an Altera Cyclone device with 20 kLogic Elements (LE), 64 blocks of internal 4 kBit RAM and two phase-locked-loops (PLLs). Furthermore, an external 1-MByte SRAM is mounted together with the FPGA.

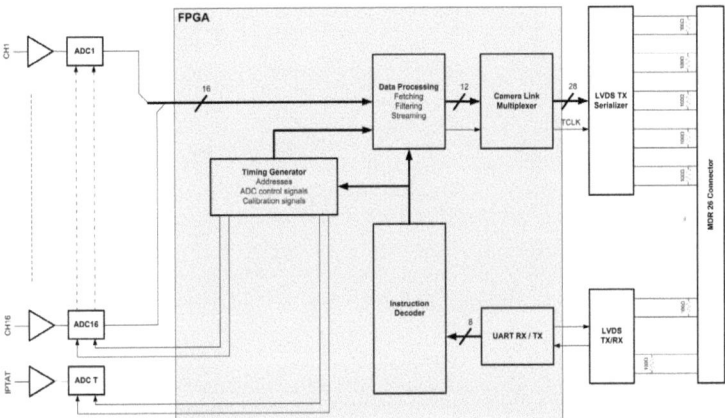

Figure 2.11 : Architecture of interface - FPGA board. The output of the FPGA (28 bits) meets the CameraLink bit assignment.

The processed data is then sent from the FPGA to a serializer. The 28-bit bus from the FPGA is multiplexed (7:1) on four low-voltage-differential-signaling (LVDS) channels (figure 2.12). We chose a CameraLink standard[1] as the physical layer of the transmission. The transmitted signals consists of an LVAL (line valid), an FVAL (frame valid) and a DVAL (data valid) as well as three ports of 8 bits each for the base configuration, six ports for the medium configuration and eight ports for the full configuration. The base configuration output consists of a high-speed downlink with four channels, a transmission channel of

[1] http://www.imagelabs.com

the clock, four channels for a high-speed uplink and a serial bi-directional low bitrate transmission. It can normally be accessed through a Mini-D-Ribbon-26 connector (MDR26). The base configuration supports currently a maximum clock speed of 85 MHz and can therefore achieve a payload of 2.04 Gbit/s. The medium configuration can transmit up to 4.08 Gbit/s and the full configuration is limited to 5.44 GBit/s. Both configurations require a second MDR connector.

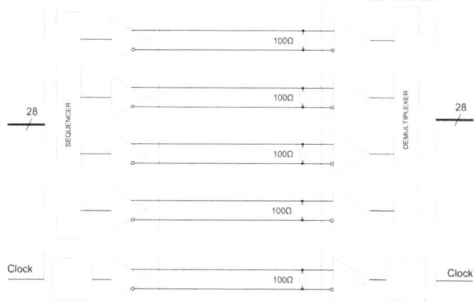

Figure 2.12 : CameraLink downlink transmitter - receiver block diagram (base configuration). The 100 Ω resistor at the receiver side is for line termination.

The CameraLink standard is adapted to the transmission of images in formats of frames and lines. It is a raw protocol and no error correction is included. The FVAL is asserted during the transmission of an image frame, the LVAL is asserted during the transmission of a line and DVAL is set to 1 when there is valid data at the output. These control signals are used by the frame grabber to trigger its buffers to the start and the end of data reception.

The architecture in figure 2.13 includes a LEONARDO frame grabber from Arvoo[2]. It organizes the data in blocks of images at both the receiver and memory level. A RISC processor is mounted on the frame grabber and allows implementing algorithms in software for image processing at a very low level. The received blocks of images are directly stored in an SDRAM (128 MB).

---
[2]http://www.arvoo.com/

2.2 System Implementation

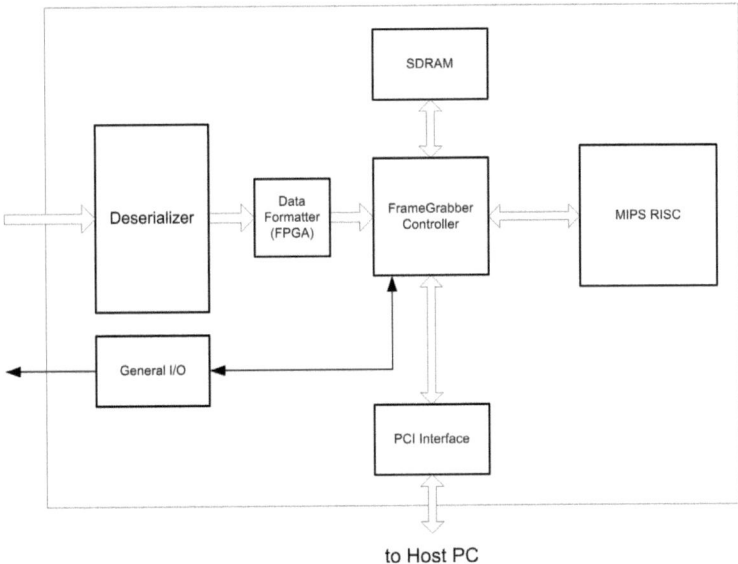

Figure 2.13 : Frame grabber architecture

The images can be transferred from the SDRAM to the host computer storage system (RAID 0) using direct memory access (DMA) through the PCI. Thus, the acquisition process does normally not interfere with the processes on the host PC.

2.2.4 Software

The control software was implemented by the Neuroengineering and BionanoTechnology group of Sergio Martinoia, at the Department of Biophysical and Electronic Engineering (DIBE) of the university of Genoa. The main application is written in C#, whereas the API functions of the LEONARDO frame grabber are provided in C++. The image processor (MIPS RISC) on the LEONARDO is programmed in C.

Even though, the processor of the host computer does not need to allocate processing time to the data acquisition itself, it is important that the data can be transferred fast enough over the PCI from the frame grabber RAM to the hard drive(s) of the computer. The disk access times are very critical and therefore a RAID 0 configuration of SAS hard drives was used. Long-term recordings of the 4096 pixels at a frame rate of 7.7 kHz provide data flows of 60 MB/s and stored files of several Gigabytes (typically 7 GB for 120 s).

For these reasons low level functions were used to access hard disk for reading/writing data, unmanaged code for data transfer from the LEONARDO board to the RAM and multithreading implementations for parallel processing. The software architecture was finally optimized in order to use the hardware features of the latest commercially available workstations which provide multi-core processing.

The software architecture provides the following features:

- Continuous recording of the entire array or of a close-up area at higher sampling frequency, continuous on-line representation of the electrical activity of the chip as a false colour map and raw data visualization of a subset of selected pixels

- Replayer mode: i.e. visualization of previous recordings that can be run forward and backward

- Stimulation configuration interface for the second generation of the APS-MEA

- Integration of off-line time domain analysis [109] i.e.:
  - Spike detection
  - Raster plot visualization
  - Average firing rate - network average firing rate

      – PSTH

      – Inter spike interval - joint inter spike interval

      – Burst detection

      – Inter burst interval

      – Network inter burst interval

- Integration of off-line wavelet domain analysis and denoising functions

- "Ad hoc" data format for high transfer rate (binary format with header)

- "Ad hoc" data format for analysis results (binary format linked to raw data)

- Raw data extraction of selected pixels in a given time interval from the original recording in our "ad hoc" format or in .mat files for Matlab.

2.2.5 Real-time Processing Units - Filters

The platform provides two blocks for real-time computation (see figure 2.2): the FPGA and the MIPS RISC image processor. One of the objectives of this work is hardware-based real-time implementation of signal processing tasks (as described in section 1.1), since this approach is more speed- and power-efficient for a high number of channels. In this section, a real-time linear filter implementation on the FPGA for the entire array is demonstrated. A more advanced hardware architecture for the computation of wavelet coefficients will be shown in a later chapter (see chapter 5).

The FPGA is a Cyclone from Altera, as indicated in chapter 2.2.3. It contains 20060 logic elements (LE) and 64 dual ports RAM blocks (256 x 18 bits each)[3]. Computing circuits can be shared by several electrodes, but registers are specific

---
[3] http://www.altera.com

to each signal. The FPGA board used as a basis for the system also provides external SRAM and SDRAM circuits. For real-time signal pre-processing of a high number of channels, the memory is one of the most critical resources. The external memory of the FPGA board does not provide adequate transfer rates (only one 32 bit-WRITE and one 32 bit-READ per sample) and therefore computation is limited by the amount of on-chip RAM blocks. We can allocate 72 bits per channel with the provided FPGA when targeting for a full array real-time filter implementation.

First, we consider a high-pass filter in order to remove APS-MEA non-idealities, such as drift and offsets. The filtering of drift effects consists in removing very low frequencies in the signal. Finite impulse response (FIR) filters need a very high order n: $n > F_s/F_c - 1$ (F_s: sampling frequency; F_c: cutoff frequency). However, infinite impulse response (IIR) filters do not need such high orders, but they are very sensitive to coefficient's precision and therefore, a floating point implementation is not an optimal solution [69]. Hence, memory constraints, and thus a limited maximum filter order, have led to the implementation of fixed-point IIR filters.

The structure of this filter is given in figure 2.14a. It requires one multiplier (α). This coefficient α determines the ratio between cut-off frequency (F_c) and sampling frequency (F_s). For optimizing the FPGA resources, we decided to limit α to $\alpha = 1 - 2^{-n}$ with n = 7, 8, 9, 10, 11 or 12. These values enable to use single bit shifters as multipliers. Possible values for α allow F_c to lie between 1 Hz and 10 Hz for sampling frequencies of 10 kHz (full array acquisition) and 20 kHz (zooming mode). Calibration acts like a step input to the filter. Thus, the filter would converge towards a new steady-state in order to cancel this new offset. To avoid the induced transient responses we update the filter state (i.e. the value of the register) with a new value after each calibration. The

2.2 System Implementation

estimation is obtained from the difference between the last filtered sample before calibration and the first raw sample after calibration.

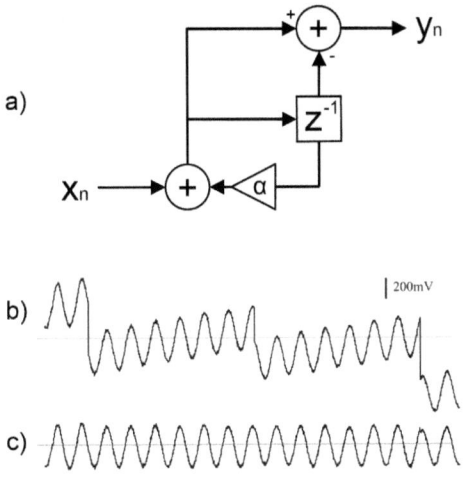

Figure 2.14 : (a) High-pass filter structure implemented in the FPGA for 4096 channels. (b) Non filtered and (c) real-time filtered acquisitions of a test signal (100 Hz sinus). The efficiency of the drift filtering and of the offset step compensation due to the amplifiers calibration are clearly demonstrated.

As the channels are electronically organized in a 256 x 16 channels matrix, registers are stored in a 256 x 320 bit RAM. We then retrieve the samples and the related registers in the same clock cycle. Filtered data are then computed in a 3-cycle pipelined unit. We implemented two units, so each of them computes 8 channels per sampling time. To compute the filter of figure 2.14a, the current implementation requires only three clock cycles and is organized in a 3-level pipelined structure. This gives the possibility to compute n samples in n+2 clock cycles. The pipelined structure is doubled in order to perform 2 x 8 samples in 8+2 cycles. Finally, data are sent and registers are updated before the next 16 channels are sampled. Figure 2.14c shows the effect of this filter on raw acquired data with calibration steps (figure 2.14b). Furthermore, the filter delays the data retrieval by 12 clock cycles (120 ns at 100 MHz). After filtering and calibration compensation, both raw and filtered samples are rearranged and sent to the computer according to the user configuration that can chose between acquisitions of raw signals only, filtered signals only or interlaced raw and filtered data. The entire logic structure in the FPGA (including instruction decoding, signal sampling, high-pass filtering and transmission) requires 2600 LEs and 23 of the 64 internal block memory, leaving 17460 LEs and 41 RAM blocks available for additional real-time pre-processing tasks (i.e. 2nd order low-pass filtering on 4096 channels).

2.3 Electrical Characterisation

2.3.1 Single Electrode with Amplifier

Individual test structures were implemented on the same ASIC in order to characterize the electrode elements.

Figure 2.15 shows the measured range of the offset compensation loop. We get an input control range from -132 mV to 20 mV for a nominal electrode bias current setting (i.e. 1 μA) and a smaller range from -90 mV to 10 mV for the minimum bias current setting. The simulated values are for 1 μA -100 mV to +20 mV and for 0.25 μA -150 mV to +20 mV, respectively.

The gain of the OTA in the electrode element was specified to 40 dB. Figure 2.16 reports on the measured gain of one pixel at different bias current settings. The value closely matches the simulated value and the pass-band of the device is 25 kHz at Ipix=1 μA and 5.5 kHz at Ipix=0.25 μA. The roll-off of the gain is due to the Miller capacitor (figure 2.10) and the decrease in cut-off frequency is induced by the smaller bias current that moves the pole at the current mirror

2.3 Electrical Characterisation

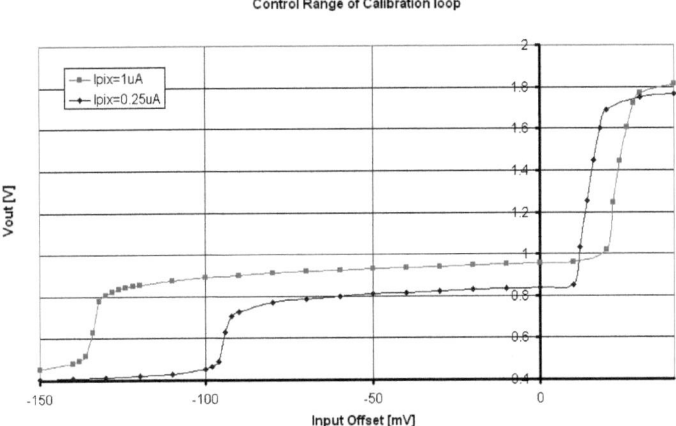

Figure 2.15 : Calibration loop control range of pixel

node down to lower frequencies.

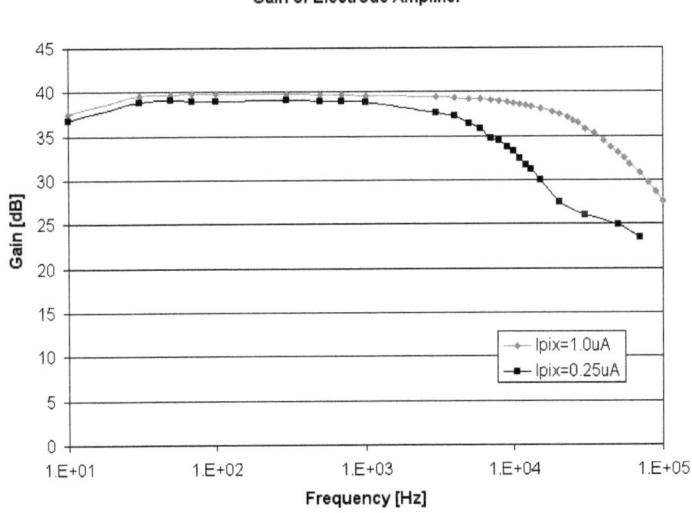

Figure 2.16 : Voltage gain of electrode amplifier in open loop.

The major contribution to the total noise power density function in the signal chain is given by the thermal noise of the electrode and noise of the pre-amplifier

Table 2.2 : Static noise measurement of pixel amplifier

Noise of pixel amplifier **without** electrode and electrolyte	Noise of pixel amplifier **with** electrode and electrolyte
$13\,\mu V_{rms}$	$22\,\mu V_{rms}$

stage [54, 110]. The noise performance of the electrode element was measured as shown in figure 2.17. The input-referred noise of wide-band oscilloscopes is usually very high. Therefore, we will have to stringently limit significant noise contribution from the oscilloscope. Two measures of precaution were taken: (i) high gain in front of the oscilloscope was inserted in order to raise the noise from the device under test (DUT) above the noise floor from the oscilloscope and (ii) the bandwidth of the oscilloscope was limited. Friis' formula [111] states that the noise from blocks following the high gain stages can be neglected. We inserted a low-noise amplifier of 13 dB after the electrode element. The noise from this additional amplifier will be masked by the 40 dB from the DUT, however it will further increase the total gain in front of the noisy oscilloscope. Assuming that the noise can be modelled as Gaussian noise the standard deviation of the noise gives the total noise power within the system bandwidth (i.e. limited by the electrode element bandwidth).

2.3.2 APS-MEA Acquisition System

An additional test row was implemented in the APS-MEA circuit. It applies a constant voltage to each column when the sensor is put in a test mode. The voltages are generated by a resistor ladder, thus, when the read-out chain of the system, including ADC, multiplexing, streaming and acquisition, is operating correctly, a falling voltage ramp should be seen from the first to the last column (figure 2.18).

2.3 Electrical Characterisation

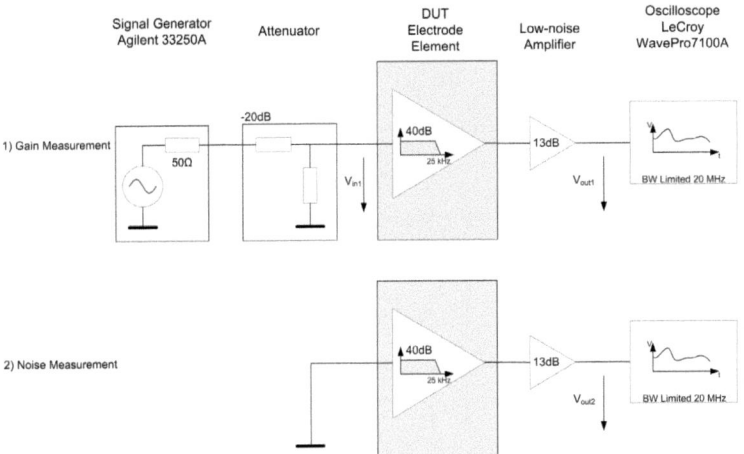

Figure 2.17 : Noise measurement setup. The signal chain to determine the noise gain (1) and the signal chain to determine the noise power of the DUT (2). Note the custom-made low-noise amplifier for better noise masking of the oscilloscope.

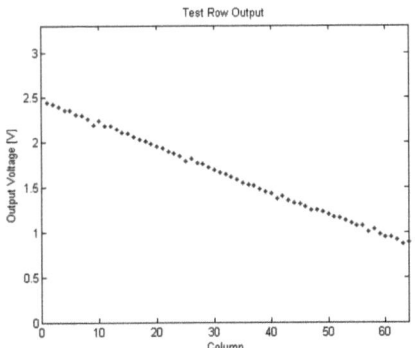

Figure 2.18 : Output of the test row

A typical output of the acquisition system is seen in figure 2.19. An artificial cardiac pulse of $100\,\mu V_{pp}$ from a signal generator was injected into the system. The raw signal output can be seen in figure 2.19a, the real-time filtered and realigned frame is shown in figure 2.19b and unfiltered waveforms of individual electrodes are depicted in figure 2.19c.

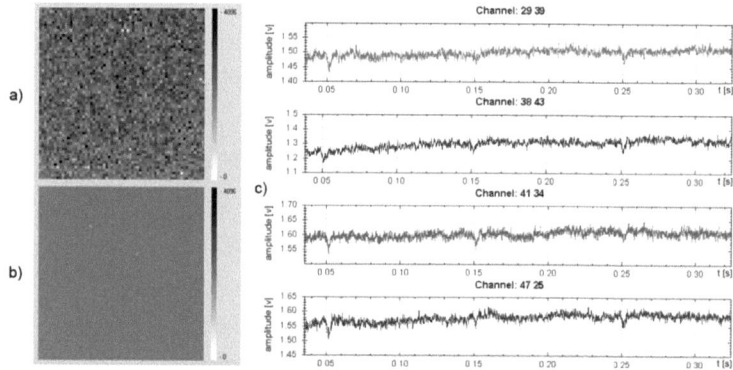

Figure 2.19 : The different data representation mode of the system are illustrated. A cardiac pulse (amplitude $100\,\mu V_{pp}$) from a signal generator is fed to an Pt-electrode immersed in a phosphate buffer solution. A raw frame is depicted in (a). The filtered output with a constant DC level (b). Four individual time waveforms reconstructed from the frame-by-frame sequence.

Figure 2.20 plots the gain of the system as a function of frequency. Finally, table 2.3 summarizes the electrical performances of the large-scale, high-resolution MEA system.

2.4 Biological Measurements

In section 2.3.2 the system was tested with respect to the specifications and required functionality. However, the final validations shall be done in neurophysiological environments that correspond to the field of application of the system. As a first intermediate step towards the complete validation we will use dissociated cardiomyocyte cultures. Since the noise floor during the dy-

2.4 Biological Measurements

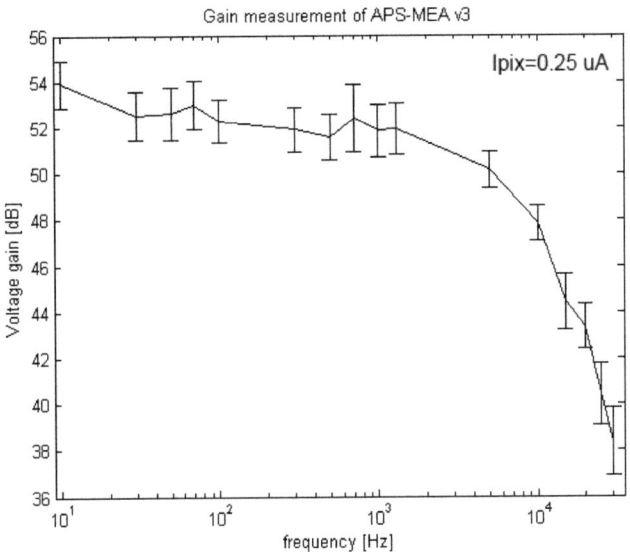

Figure 2.20 : Gain of APS-MEA acquisition system from 10 Hz to 30 kHz. The gain was measured over 128 electrodes.

namic operation of our system is in the order of $25\,\mu V_{RMS}$ (i.e. $150\,\mu V_{pp}$), the detection of the signals from cortical neurons (i.e. $\approx 20\text{-}200\,\mu V_{pp}$ measured on conventional MEAs) seems to be difficult without knowing the precise nature of waveforms recorded with high-density MEAs. Therefore, we start measuring signals from cardiomyocytes as their expected amplitudes are in the order of millivolts. Furthermore, their action potentials last longer ($\approx 20\,\text{ms}$), which makes it easier to detect them with a sampling rate of 8 - 10 kHz. After having successfully measured biological relevant signals from cardiomyocytes, we will move on to the measurement of neuronal cells. A perfect validation would be performed only if we matched the extracellular recordings from our system with intracellular measurements. However, this is not currently feasible and therefore we will need to rely on the experience from state-of-the-art MEA measurements (i.e. in terms of signal shape, spike / burst durations, etc.) and

Table 2.3 : Summary of characteristics of the high-density APS-MEA system

Number of Electrodes	64 x 64
Electrode Size	21 μm x 21 μm
Electrode Pitch	42 μm
Active Area	2.67 mm x 2.67 mm
Chip Size including Pads	5.5 mm x 5.3 mm
Supply Voltage	3.3 V
Overall Power Consumption of Chip	132 mW
ADC Resolution	12 bit
Input-Referred Noise (static)	13 μV_{rms}
Input-Referred Noise (dynamic)	26 μV_{rms}
Input Signal Range	100 μV_{pp} - 2 mV_{pp}
Minimum Amplification	55 dB
Minimum Frame Rate (4096 electrodes)	8 kHz
Maximum Frame Rate (64 electrodes)	125 kHz
Total Data Throughput in Full Mode	500 Mbit/s

correlations between standard fluorescent imaging methods and electrophysiological activity of cultures under test. Furthermore, we will apply a neuroactive drug, cyclothiazide (CTZ), that has a known impact on activity of neuronal cultures.

2.4.1 Reference Culture on Passive High-Density MEAs

Large-scale MEA systems with metallic electrode densities of over 500 electrodes per square millimeter, as we implemented in our MEA system, have not been reported yet in literature. The nature of the signals that we can potentially get

2.4 Biological Measurements

is unknown *a priori*. The reasons are twofold: (i) expected signal contributions from one neuron to several electrodes and (ii) unknown coupling effects from the cells and the heterogeneous surface of the commercially fabricated APS-MEAs. We tried to take into account the former reason by implementing passive high-density MEAs using thin-film microfabrication technology. These MEAs were designed by Luca Berdondini at the SAMLAB. The form factor of these MEAs was designed to fit into the commercially available recording system from Multichannelsystems in Reutlingen, Germany. Titanium and Platinum were deposited on either a Pyrex© or silicon substrate. Electrode diameters were 22 μm or 30 μm, with separations of 10 μm or 20 μm between electrodes, respectively [103]. An example of dissociated cortical neurons from rat, plated on a passive high-density MEA is shown in figure 2.21. Signals, shapes, burst and spike duration of a spontaneous activity are similar to values obtained with conventional MEAs [112]. Spike durations from 1 to 3 ms can be observed and burst lengths up to several hundreds of milliseconds can occur. Typical spike shapes are seen on the right of figure 2.21.

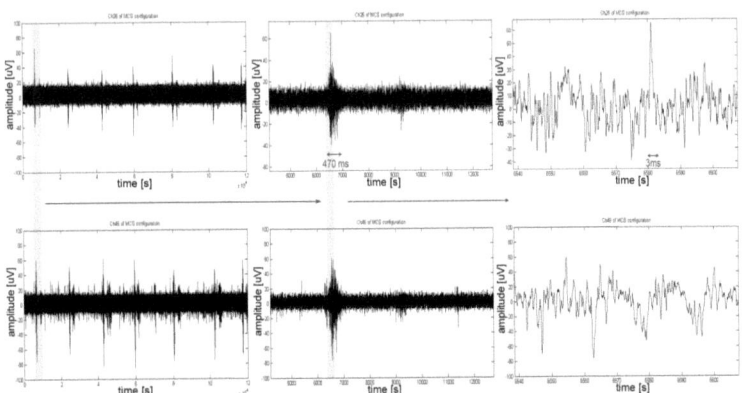

Figure 2.21 : Spontaneous neurophysiological signals from two channels of a passive high-density MEA (electrode diameter 22 μm, separation 10 μm). The cells are from embryonic rat (E18) at an age of 20 DIV [103].

2.4.2 Dissociated Cardiomyocytes Cultures

The biological protocol was developed at DIBE, Genova. The first step in the validation process involves biological tests with cardiomyocytes cell cultures. For this purpose, ventricular cardiomyocytes were obtained from Sprague Dawley rats at embryonic day 14-15 (E14-15) and processed as follows. Heart tissue fragments, once isolated and washed in Ca^{2+} and Mg^{2+} free Hank's balanced salt solution, were incubated with 0.05% Trypsin (E14-15) or 0.125% (E18) at 37°C and exposed at two or three cycles of enzymatic digestion. The supernatant containing dispersed cardiac cells was removed after each cycle and placed into a conical centrifuge tube with a solution of nutrient medium (5% FCS, 1 mM L-Glutamine, 1% Pen-Strepto, DMEM-F12). After the last cycle of Trypsin, cells were mechanically separated in nutrient medium by repeated pipetting and centrifuged for 5-8 minutes at 1000 r/min. In order to separate cardiac fibroblasts from ventricular cardiomyocytes, the cells were pre-plated in a Petri dish, without any treatment of the adhesion factor, with nutrient medium and stored for 1 h at 37°C in a humidified standard incubator. Finally, ventricular cardiomyocytes were collected, centrifuged, diluted in DMEM-F12 medium, 4%FBS, 2%HS, 1% Pen-Strepto and plated onto the pre-coated (Laminin solution at 0.01 mg/ml overnight) APS-MEAs. Three days later, the medium was replaced with a serum-free medium composed by DMEM-F12 supplemented with 1% N2.

Cardiomyocytes cultures grown on APS-MEAs showed good viability over the observed period of two weeks and their behavior was similar to reference cultures simultaneously plated on standard MEA devices. Even if the cells showed a spontaneously contractile behavior after about 48 hours *in vitro*, we used cultures at 4, 5 and 7 DIV in the experiments in order to observe the propagation

2.4 Biological Measurements

of large and clear electrophysiological signals. An example of acquired full frame activity is illustrated in figure 2.22 by 2d- and 3d-plots captured at three distinct acquisition times. Additionally, three electrode signals acquired at locations indicated on the 2d-plots of figure 2.22c are reported as amplitude-time graphs in figure 2.22b. These results demonstrate how the high spatio-temporal resolution of our system enables detailed observation of the cardiac pulse propagating wave.

The functionality of the zoom mode is demonstrated by the results reported in figure 2.23 where an area of 384 electrodes was addressed. In this case, the temporal resolution is increased down to $18\,\mu$s.

2.4.3 Dissociated Cortical Cultures

The biological protocol was developed at DIBE, Genova and at INSERM, Bordeaux. The APS-MEA was coated with $100\,\mu$l Poly-D-Lysine (0.1 mg/ml) after sterilization in Ethanol (70 %) and incubated overnight at 37°C, 5 % CO_2 and 100 % humidity. Poly-D-Lysine was removed and $50\,\mu$l of Laminin (1 mg/ml) were added on the electrodes. Then, cerebral cortical pieces from fetal rats at E18 were incubated with 1.5 ml of Trypsin for 25 minutes. Cortex pieces were dissociated by enzymatic digestion. The Trypsin was removed, 10 ml of the nutrient medium DMEM-P-Glutamax I was added and the solution of cells was centrifuged at 1000 r/min. After removal of the media, the Trypsin was neutralized by adding $100\,\mu$l STI (Soybean Trypsin Inhibitor) with 3 ml of DMEM-P-Glutamax I. The pieces were dissociated with a glass Pasteur pipette. 10 ml of DMEM-P-Glutamax was added again and the cells were centrifuged for 5 minutes with 1000 r/min at room temperature. $500\,\mu$l of Trypsin and $500\,\mu$l of DMEM are added after having removed the supernatant. After 1 minute the

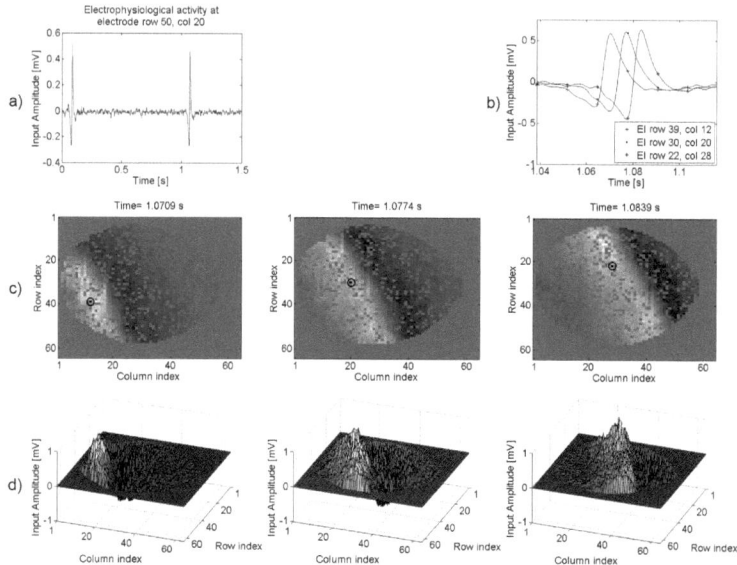

Figure 2.22 : Electrophysiological activity recordings of rat embryo cardiomyocytes at a frame rate of 8 kHz. (a) Recording from the electrode at row 50, column 20 during 1.5 seconds. (b) Time sequences from three different electrodes ((i) Electrode at row 39, column 12, (ii) Electrode at row 30, column 20 and (iii) Electrode at row 22, column 28) are shown. The propagation delay between the distant electrodes can be clearly seen. (c) 2d-representation from our acquisition software at different time instants and showing the entire array (4096 electrodes). The amplitudes are normalized on a gray-scale, ranging from the minimum value in black to the maximum value in white. The solid circles on the three graphics correspond to three specific electrodes whose signals are also shown on (b). The gray surrounding area indicates the region on the APS-MEA that was covered with glue during packaging. (d) A 3d-representation of cardiac action potentials propagating over the array.

Trypsin was again neutralized by 100 μl STI. Then, 50 μl DNAse was added with 10 ml DMEM, again centrifuged for 5 minutes with 1000 r/min at room temperature and the supernatant was removed. The cells were diluted in 2 ml of Neurobasal Medium (NBM) supplemented with 2 % of B27, 1 % of Glutamax and 1 % of Antibiotic (Penicillin/Streptomycin). Thereafter, the cells were counted using trypan blue and diluted to $2.5 \cdot 10^6$ cells/ml. The Laminin was then removed from the APS-MEA and 100 μl ($\approx$ 250'000 cells) of the cell suspension was dropped on the array surface. The preparations were incubated

2.4 Biological Measurements

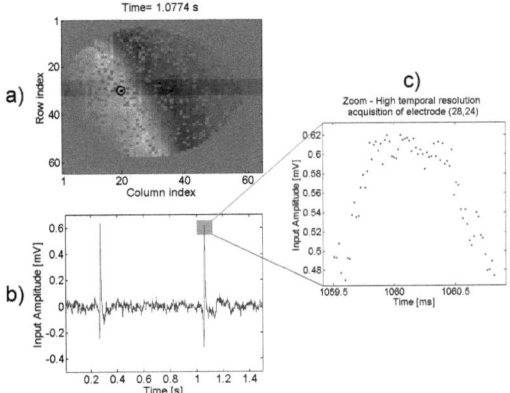

Figure 2.23: (a) The highlighted subset of the array selects 384 electrodes used for an increased temporal resolution acquisition. (b) Signal of one selected electrode (indicated in (a)) at a frame rate of 55 kHz. (c) The enhanced temporal resolution in the zoom mode can be achieved by reducing the number of addressed electrodes used for a fine analysis of the micro-scale propagation effects.

with the supplemented NBM and changed twice a week.

Figure 2.24 shows a frame sequence of a burst of an acquisition from a dissociated neuronal network. The cells are extracted from a rat at embryonic day 18 (E18). The cell density was $2.5 \cdot 10^6$ cells/ml. An example of individually extracted electrodes from the same preparation is shown in figure 2.25.

2.4.4 Dissociated Hippocampal Cultures

Primary cultures were obtained from brain tissue of Sprague Dawley rats at E18. Briefly, embryos were removed and dissected under sterile conditions. Hippocampi were separated from cortex and dissociated by enzymatic digestion in Trypsin 0.125 % - 20 min at 37°C- and finally triturated with a fire-polished Pasteur pipette. Dissociated neurons were plated onto Poly-D-Lysine- and Laminin-coated APS-MEAs. In order to optimize cell cultures' growing several conditions of plating were tried. To cover the whole area of $2.67 \times 2.67 \, mm^2$

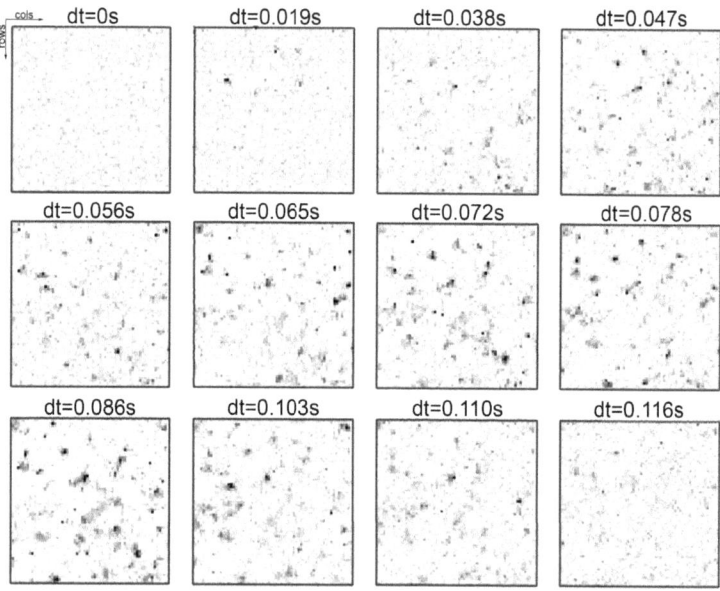

Figure 2.24 : Burst sequence of a dissociated cortical culture (from rat, E18) at 23 DIVs. Courtesy of INSERM, Bordeaux.

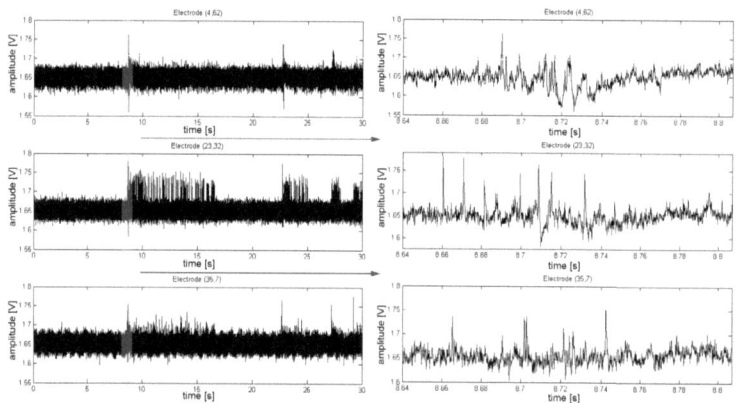

Figure 2.25 : Single channel activity of the culture from figure 2.24 during 30 seconds. The amplitude is referred to the output of the amplifiers (52 dB).

2.4 Biological Measurements

drops of variable volumes of cell suspension from 25 to 40 μl were used in which cell concentration was calculated between 250 - 600 cells/μl. One hour later, when cells adhered to the substrate, 1 ml of medium was added in each device. The cells were incubated with 1 % Glutamax, 2 % B-27 supplemented Neurobasal Medium (Invitrogen), in a humidified atmosphere 5 % CO_2, 95 % air at 37°C. The 50 % of the medium was changed every week. No antimitotic drug was used, because application of serum-free medium limits the growth of non-neuronal cells, necessary for the development of a healthy preparation.

Figure 2.26 shows a frame sequence of a burst of an acquisition from a dissociated neuronal network. The cells are extracted from a rat at embryonic day 18 (E18). An example of individually extracted electrodes from the same preparation is shown in figure 2.27.

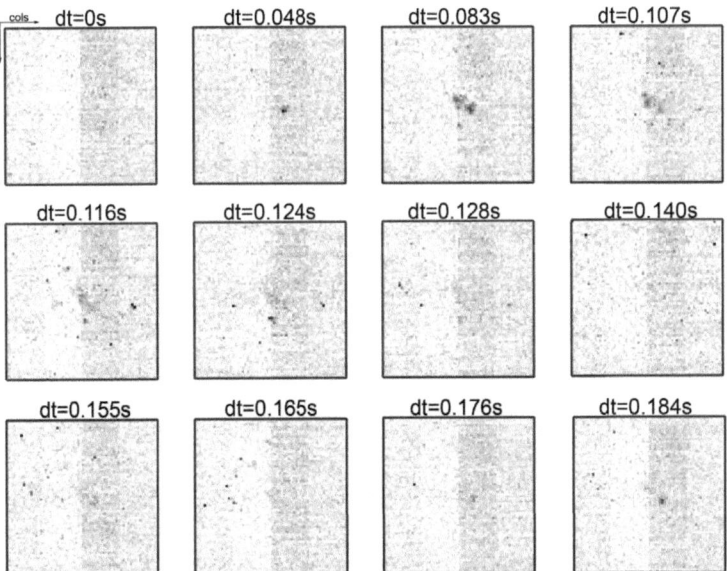

Figure 2.26 : Burst sequence of a dissociated low-density hippocampal culture at 14 DIVs. Courtesy of DIBE, Genova.

An alternative way to validate our system is the matching of the topology of

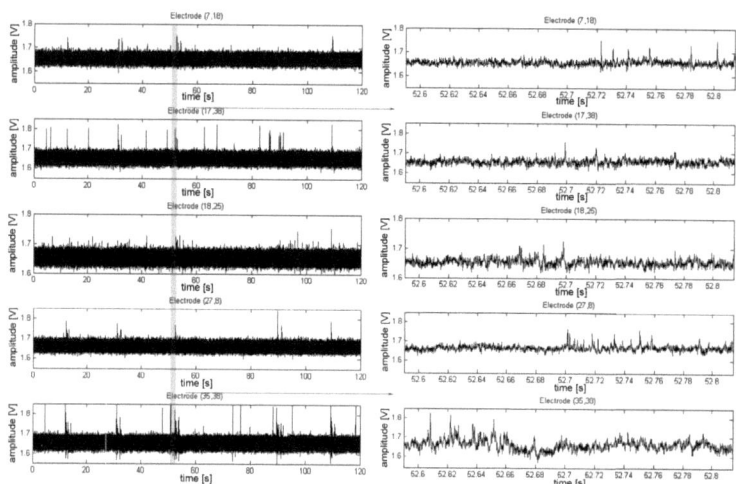

Figure 2.27 : Single channel activity of the culture from figure 2.26 during 30 seconds. The amplitude is referred to the output of the amplifiers (52 dB).

a neuronal network with its electrophysiological activity. The electrodes that record activity are likely to be physically close to active neurons. Therefore, a correlation between the physical structure of the network (i.e. obtained by staining and imaging) and the recorded activity should be apparent.

Figure 2.28 is an attempt of showing this correlation on a low density (i.e. 8000 cells/25 μl) hippocampal cell culture from rat (14 DIV). The upper row corresponds to the fluorescent image where NeuN (neuronal nuclei) was used to stain the nucleus of the neuronal cells. The lower row of the figure shows an integrated activity metric for a 1 s interval including a burst sequence. The metric was computed by using the kurtosis [113], a higher order statistics that measures the peakedness of the probability distribution of a random variable. As expected the overlapped images in figure 2.28 indicate a considerable correspondence of the two complementary representations. However, the correlation is not maximal yet. This is due to a limited noise performance of the acquisition system as well as not yet optimal cell culture conditions, such as culture homogeneity, cell adhesion and cell density.

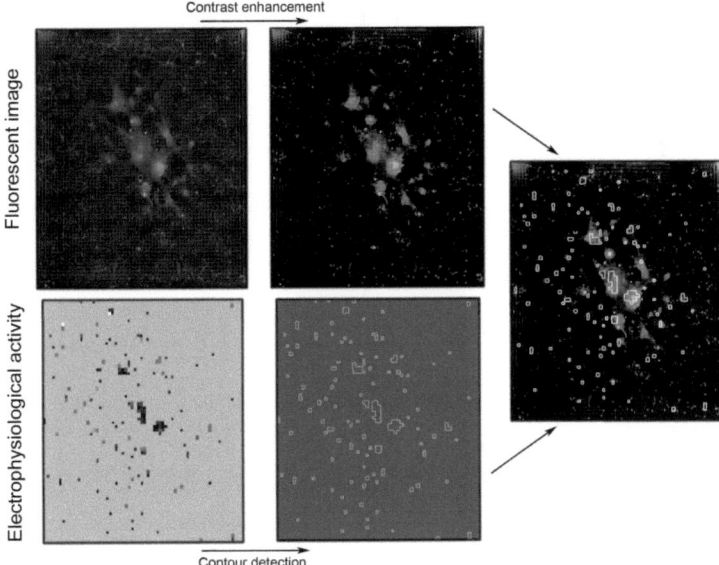

Figure 2.28 : Overlap of an activity image computed from electrophysiological acquisitions (see figure 2.26) and a fluorescent image (MAP and NEUN staining) of the culture [fluorescent images and activity recording provided by DIBE, Genova]. The activity contours (bottom) match the biological cells (top) to a great extent. The dimensions of the array are 2.67 mm x 2.67 mm.

It is known that CTZ can induce eptileptic activity [114] in an *in vitro* network of hippocampal neurons. Thus, one further validation technique of our platform is to record a reference activity and then induce an increase in bursting activity by adding CTZ. The measured effect is shown by comparing two raster plots before and after administration of CTZ (figure 2.29). A raster plot is a map that shows all spike events of all channels as a function of time.

2.5 Conclusion

The functional, electrical and biological performances of a new large-scale high-resolution microelectrode array platform were demonstrated. The system is

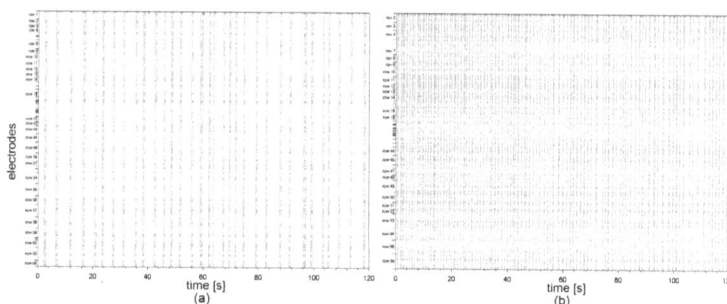

Figure 2.29 : Raster plot of hippocampal neuronal cells (DIV27) before (a) and after (b) administration of CTZ. Courtesy of DIBE, Genova.

based on concepts stemming from the image acquisition and processing field. It consists of a CMOS-based APS circuit where the light-sensitive elements were replaced by metallic electrodes. The size of the electrodes ($21\,\mu$m x $21\,\mu$m) and the distance between electrodes ($21\,\mu$m) enable monitoring dissociated neuronal networks at both network and cellular level. The platform can continuously record from 4096 electrodes at the time. Whole frame sampling rates at $8\,$kHz generate an overall data rate of about $500\,$Mbit/s. One of the bottlenecks of such a high number of electrodes is the processing time that has to be allocated to analyze the data. Current systems only provide computer-based processing. For a small number of electrodes (≈ 60 - 100) the performance of a host computer might be sufficient for on-line or even real-time processing, however a speed-efficient analysis throughput is unrealistic for several thousands of electrodes. Extensive hardware resources, such as FPGA and RISC units, underline the strong emphasis to enable real-time processing of the data.

The different blocks of the acquisition system were discussed: (i) the APS-MEA that integrates the front-end electronics (i.e. switches, amplifier), (ii) the interface board including ADCs, FPGA and data multiplexer, (iii) the acquisition board on the host computer and (iv) the software that controls the acquisition

2.5 Conclusion

and allows off-line signal processing. Furthermore, real-time signal processing was implemented in hardware on the FPGA. In that sense, a low-order high-pass filter that processes all 4096 signals was designed in order to remove non-idealities of the APS-MEA and in order to significantly enhance the signal quality. Further techniques for real-time signal enhancement and concepts of hardware-based real-time data analysis will be discussed in chapter 3 and 5.

The second dedicated hardware resource, the MIPS RISC image processor on the acquisition board, has not been used yet. The present work focused on hardware-based implementation and therefore, we omitted a detailed discussion of all potential applications of the image processor.

After the detailed discussion of all the architectural blocks, the performance of the system was evaluated. Individual test structures of single electrode elements allowed the testing of the electronics under static and controlled conditions. Critical parameters, such as overall gain and system noise were measured and compared with expected (simulated) values.

The final aim is to use the system under a wide range of experimental conditions for investigating relations between local phenomena in electrogenic cells. Such experiments can enable studies of propagation effects, synaptic changes and network characteristics, but also overall dynamics (e.g. firing rate, network bursting), information processing mechanisms (i.e. population code) and network plasticity. Therefore, the system was validated under biological conditions and a series of tests on dissociated cultures of cardiomyocyte, hippocampal and cortical cells from rat were performed. Additionally, a basic neuroactive compound (CTZ) was used to induce chemical stimulation into the network.

In conclusion, the system showed adequate functional characteristics with respect to noise, system bandwidth and processing speed enabling continuous recordings of both local and global levels of activity of cardiomyocyte, hippocampal and cortical cultures.

Contributions

To the best of our knowledge, the following contributions can be considered as original:

- Concept and specification of system blocks for a large-scale, high-density microelectrode array acquisition platform including
  - Architecture of new APS-MEA application specific integrated circuit
  - Architecture of a scalable signal processing and acquisition hardware
  - Management of data (i.e. image-based protocol)
- Implementation and validation of a large-scale, high-density microelectrode array acquisition system for neurophysiological signals

Further potential research

- Deep submicron CMOS technologies enable the implementation of more sophisticated circuitry, such as correlated double sampling or chopper-based methods, to reduce more efficiently noise and drifts in the electrode element. For constant electrode size and separation more available silicon area will allow investigating different architectural methods for the electrode element implementation in order to enhance the electrical characteristics of the sensor (i.e. noise, offset, etc.)

- Integration (i.e. on-chip) of analog-to-digital conversion and digital signal processing units (i.e. filters) for enhancement and analysis of the biological data

Chapter 3

Real-Time Signal Processing for High-Density MEA Systems

In the previous chapter of this thesis we discussed a high-density APS-MEA platform that records signals from 4096 electrodes. This high-density CMOS-MEA system [115] produces data streams in the order of several hundreds of Megabits. Real-time processing becomes mandatory for convenient experimentation (i.e. on-line observation and selection of important features) and for managing the substantial amount of data generated in long-term recording sessions. The high level of integration on the chip entails the need of pre-processing of the sensor output for managing the noise, mismatches and other non-idealities (see chapter 2.2.5). The low SNR compromises the use of simple detection methods, such as threshold-based spike detection. More advanced signal processing techniques for signal enhancement and detection are available but often lead to a significant increase in computational power. Brute force processing of the data within a computer grid is always possible, however is not an appropriate way to limit the complexity and usability in a biological laboratory. Moreover, to fully exploit a high-resolution methodology allowing multi-dimensional access to the network activity, ranging from the whole network to the cellular/sub-cellular levels, it is required to further develop signal processing methods in order to extract the relevant signals (e.g. spikes and bursts) from the large amount of acquired electrophysiological data.

An alternative solution to handle the data is the implementation of a large part of the necessary operations in hardware. However, hardware-based analysis tends to have limited flexibility for extensions to tasks that it was not originally developed for. It is therefore a challenge to overcome as many constraints as possible to allow a large set of potential analysis methods. We need a concept that implements the conventional analysis functions for MEA signal processing and that enables further extension to new processing ideas for high-resolution MEA recordings.

In this chapter of the thesis we will demonstrate a flexible signal processing framework. This framework allows conventional MEA signal processing of a large number of electrodes by extending wavelet-based analysis techniques for signal enhancement, spike detection and sorting and, subsequently, by implementing a processing unit for the hardware integration of these tasks (described in chapter 5). These concepts are based on a channel-by-channel processing of each electrode. Wavelets offer solutions to many potential problems in signal processing and estimation and obey a set of properties that lead to efficient algorithms implementable in both hardware and software. They have already been used for several years in the field of neuroelectrophysiology. Spike detection and classification from (multi-)electrode systems in *in vivo* experiments were implemented [87, 89, 116, 117]. Since the fundamental unit of neuroelectrophysiology is a duration-limited action potential generated by an individual neuron, it is intuitive that wavelet signal theory potentially performs better than Fourier-based signal processing where the basis functions are made of infinite duration sinusoidal/cosinusoidal signals. The best representation of a signal in a transformed domain, i.e. compaction and decorrelation of the signal into a sparse set of coefficients, can be obtained if we select a basis that consist of signals that look as similar as possible to the signal to be transformed. A typical shape of an action potential recorded by an extracellular technique was shown in figure 1.5 of chapter 1. Many different wavelet systems that were implemented in filter banks have been proposed [118, 119]. In that sense, Symlet5, Symlet7 and Biorthogonal5 are often the filters of choice in the field of neurophysiological signals [108].

3.1 State-of-the-Art and Motivation

Low-density MEAs employ amplification and band-pass filtering blocks to enhance the signal quality [54, 55]. In these MEA systems simple analogue pre-processing is usually sufficient to reject low frequency interference as well as high frequency noise. However, CMOS-based high-density MEA circuits aim at integrating thousands of electrodes on a single chip and therefore they can not benefit from relatively large allocated areas for the implementation of typical high- and low-pass filtering blocks next to each electrode. Furthermore, such high-density MEAs are also critical in terms of noise. Noise-optimized amplifiers close to the electrodes, i.e. large transistor sizes and increased current consumptions, are better for decreasing the noise, but these constraints are incompatible with the high integration. A trade-off has to be defined at the circuit design level requiring, in parallel, the use of sophisticated denoising techniques in order to detect relevant biophysical information from low SNR signals. Moreover, high-density MEAs use alternative circuit architectures and biasing schemes to decouple DC signal components [61, 115]. These circuits can introduce additional mismatch, offset and drift components in the output signal. Therefore, the lack of space and the need for sophisticated detection techniques entail that the pre-processing has to be moved from the analogue to the digital world and, hence, requires additional computational resources to handle this task. As already mentioned in chapter 1 signal processing methods in the field of microelectrode arrays have been developed since the sixties [82, 120]. The fundamental information of a neuronal network are spikes, or equivalently, action potentials of neurons. Thus, each signal from an electrode is analysed for spike events and various patterns and relationships, such as cross-correlation, post-stimulus timing histogram, etc., between neurons are then computed [94, 121–123].

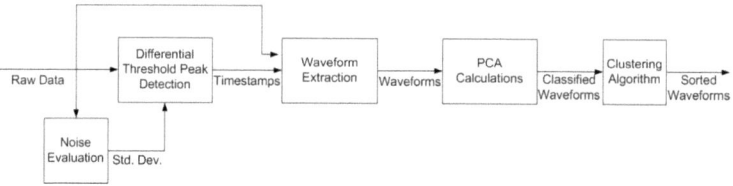

Figure 3.1 : Flow chart of conventional analysis method for extracellular electrophysiological signals acquired on MEAs.

One microelectrode can record from several neurons. Besides the detection of spikes, it is therefore critical to discriminate between signals of different neurons acquired at one electrode. This increases the number of neurons that can be monitored. Therefore, conventional methods for extracting and sorting neuronal events from continuously recorded MEA signals consist essentially in the identification of the intrinsic signal spikes and in clustering the events into groups. In this way, the different neuronal contributions sensed at each electrode site and arising from the surrounding neuronal populations are identified and separated [71, 90, 124, 125]. A good review of the methods for spike detection and sorting can be found in [83].

Different approaches exist to build spike detectors. For example, Obeid [84]

3.1 State-of-the-Art and Motivation

evaluated different types of threshold-, energy- and matched filter-based spike detectors. Also for the sorting, several numerical tools were reported [83, 126, 127] and many clustering algorithms presented [128]. PCA (principal component analysis) is the method of choice since many years [91, 129] and commercialized tools are available (e.g. Offline Sorter from Plexon[1]). In general, the performances of these methods are affected by the initial event detection step and the analysis is performed off-line. In order to illustrate the conventional approach, here we consider a prototypical analysis method composed by i) a differential threshold-based spike detection [130] and ii) a spike classification based on PCA as performed on the commercially available Offline Sorter. The flow chart of the analysis steps is reported in figure 3.1. We also point out that PCA is very costly in terms of mathematical operations since it involves the computation of cross-correlations and eigenvectors of many vectors and large matrices, respectively [77].

The analysis of the raw data acquired on MEA systems featuring a small number of read-out electrodes, is typically performed on a host computer after external [131] or on-chip [58, 59, 132] analogue signal conditioning, i.e. amplification, filtering and analogue-to-digital conversion. However, analysis using methods described above (i.e. PCA) is a computationally expensive task. Relatively low data rates from conventional MEAs allow handling the data in real-time or on-line modes [133, 134]. To enhance the system functionality, the integration of fundamental processing tasks for neurophysiological applications, such as spike detection by thresholding on FPGA [64] and the implementation of individual hardware-based processing blocks for denoising, detection and compression of signals from a low number of channels has been reported [66–68]. However, an entire processing framework for large-scale (i.e. large number of electrodes) and

---

[1] http://www.plexoninc.com

high-resolution MEA acquisition systems enabling on-line or real-time analysis has not been proposed yet. Therefore, we will focus on a *wavelet-based* signal processing framework that allows:

- Denoising low SNR signals

- Performing spike detection and spike sorting

- Implementing real-time functions conveniently in hardware

- Enabling spatially high-resolution (i.e. image) signal processing for new feature extraction of electrogenic cell networks

3.2 Denoising

Donoho's pioneering work [135] suggested that the wavelet transform can be used for efficient denoising of finite-support signals. A sparse set of relatively large wavelet coefficients can represent a large class of signals (see appendix A). Temporal noise is spread out to many but small coefficients in the wavelet domain. For illustrative purposes, figure 3.2 compares the denoising results of an experimental high-density MEA signal (cortical culture, DIV 23) from one electrode at a given low SNR obtained by simple linear filtering and by denoising in the wavelet domain. A cut-off frequency at 2 kHz and a 30-order finite impulse response filter (FIR) were used. The sampling rate of the signals is 7.7 kHz.

Donoho's denoising algorithm includes following steps:

- Compute orthogonal discrete wavelet transform (DWT) $\rightarrow c_k, d_k$

- Soft- or hard thresholding of the wavelet coefficients

3.2 Denoising

– Universal threshold:
$$\tau = K \cdot \sigma \cdot \sqrt{2 \cdot \ln N} \quad (3.1)$$

σ is the standard deviation of the finest scale wavelet coefficients [135], N is the number of samples and $K \in [1,2]$; σ can be robustly estimated for Gaussian white noise:
$$\sigma = \frac{median(|d_k|)}{0.6745} \quad (3.2)$$
where d_k corresponds to the finest scale detail

– Soft-threshold
$$\tilde{d}_k = \begin{cases} 0 & \text{for } |d_k| < \tau \\ sign(d_k) \cdot (|d_k| - \tau) & \text{for } |d_k| \geq \tau \end{cases} \quad (3.3)$$

– Hard-threshold
$$\tilde{d}_k = \begin{cases} 0 & \text{for } |d_k| < \tau \\ d_k & \text{for } |d_k| \geq \tau \end{cases} \quad (3.4)$$

- Compute corresponding inverse wavelet transform (IDWT) from c_k and $\tilde{d}_k$

The method is schematically shown in figure 3.3 where, for denoising applications, the block in the center implements the thresholding.

The classical denoising algorithm that was proposed by Donoho is based on the orthogonal DWT. However, as seen in appendix A the orthogonal DWT is not shift-invariant. Hence, the denoising gives different performances depending on the position (i.e. shift) of the spike. Donoho's concept was extended to shift-invariant redundant wavelet transforms [136]. Simulations illustrate the performance of both orthogonal DWT-based and shift-invariant DWT (SIDWT)-based denoising applied to typical models for extracellular MEA signals. A spike model from [44] was used to get the ideal signal f_{ideal}, normalized to 1,

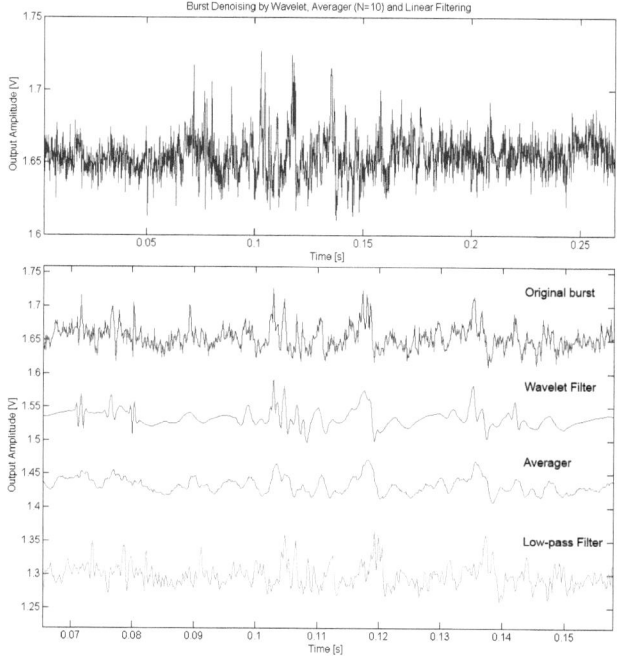

Figure 3.2 : Original burst signal from a cortical culture (DIV 23) measured at one electrode. Wavelet-denoised signal according to Donoho's method, averager of order 10 and low-pass filtered signal (30th order).

Figure 3.3 : Generic processing chain for transform-based operations

and Gaussian white noise $N(0, \sigma_n)$ (figure 3.4a) was then added. The input SNR is defined as

$$SNR_{in} = 20 \cdot \log_{10}\left(\frac{1}{3 \cdot \sigma_n}\right) \quad (3.5)$$

The output root-mean-square-error

3.2 Denoising

$$e_{rms} = \sqrt{\frac{1}{N} \cdot \sum_{k=1}^{N} \left[\hat{f}_{out}(k) - f_{ideal} \right]^2} \qquad (3.6)$$

between the original (i.e. simulated) and denoised signal $\hat{f}_{out}$ was computed. The output SNR is then defined as

$$SNR_{out} = 20 \cdot log_{10} \left(\frac{1}{3 \cdot e_{rms}} \right) \qquad (3.7)$$

It is computed for different input noise power and random shifts in time (figure 3.4). Symlet5 is used as mother wavelet for the DWT.

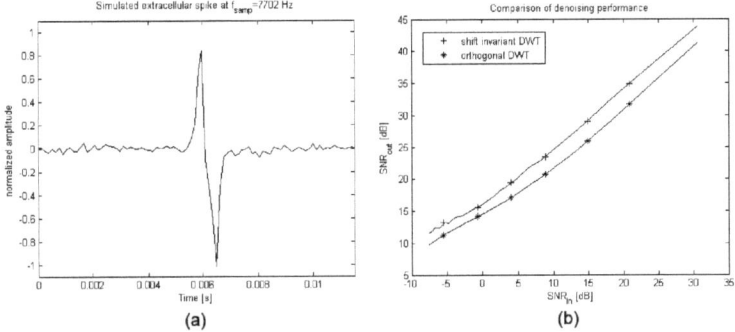

Figure 3.4 : Denoising by thresholding in the wavelet domain using the shift-invariant DWT and the orthogonal (classical) DWT. Symlet5 is used as the mother wavelet.

The output SNR of the SIDWT is about 3 dB higher with respect to the orthogonal DWT. The redundant representation of the signal in the SIDWT space increases the performance of the denoising operation. This improvement is obtained at the expense of a higher computational complexity of the SIDWT. In the orthogonal DWT the noise is uncorrelated between scales, however in the shift-invariant non-orthogonal case there is redundancy, thus noise from different scales is also correlated to some extent. Therefore, the universal threshold from equation 3.1 might even be too conservative in the SIDWT case. A further optimisation of the threshold τ depending on the application (i.e. signal waveform) can be performed for both SIDWT and DWT.

Figure 3.5 evaluates the denoising performance of the Symlet3 and the Symlet5 wavelets applied on a template spike obtained from the averaging of spikes of the APS-MEA. Symlet5 performs slightly better than the Symlet3. However, the filter order for Symlet5 is 10 whereas it is 6 for Symlet3. The lower order wavelet is obviously preferred for real-time implementation.

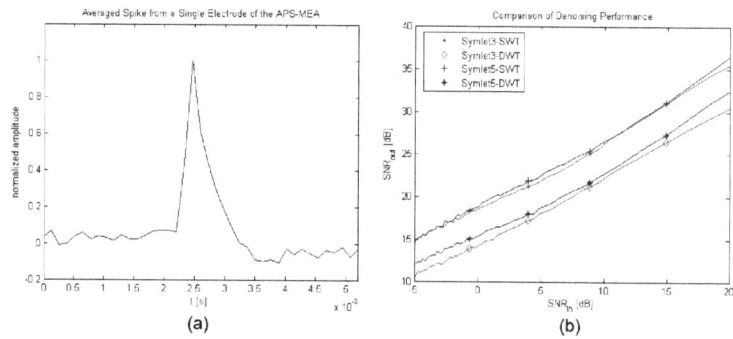

Figure 3.5 : Spike template (averaged) from a APS-MEA single electrode measurement from a dissociated cortical culture from rat (DIV 23) (a). Denoising performance of a simulated spike train using the template from (a) with variable Gaussian white noise (b).

3.3 Spike Detection

A valuable alternative to the conventional processing of neurophysiological signals is to perform spike detection [87, 89] and classification [87, 92, 137, 137, 138] with algorithms based on the wavelet transform. This enables efficient non-linear detection methods [72, 73, 135]. Wavelet-based detection is particularly well suited for low SNR signals [85, 88] and it is closely related to the denoising discussed in the previous section. Furthermore, wavelet-based detection can

3.3 Spike Detection

successfully approximate optimal linear matched filters [70] without an *a priori* knowledge of target waveforms. The operator block in figure 3.3 can implement simple thresholding as described in section 3.2 or more advanced nonlinear techniques, such as combining wavelet coefficients from different scales [88]. The main difference with respect to denoising is that the inverse wavelet transform is omitted since the occurrence of a spike can be determined in the wavelet domain.

The different scales of wavelet decomposition allow observing a signal at different levels of details. Sharp edges in a signal will be represented by higher coefficients at lower scales (i.e. higher resolution) whereas smoother signal contributions are more likely to be found at higher scales (i.e. lower resolution). This is in contrast to the Fourier transform of a sharp transition where any time information is blurred due to the duality of time and frequency. The advantage of the wavelet based detection is that multiple thresholds can be introduced in order to allow the precise selection of spikes according to their waveforms characteristics (see also section 3.4).

A simple algorithm for wavelet-based spike detection includes following steps:

- Compute wavelet transform
- Apply threshold τ_j (universal threshold according to equation 3.1, empirical threshold or adaptive threshold [139]) to each scale
- Select the wavelet transform modulus maxima (WTMM)
- Merge information from all scales, e.g. consider only contiguous regions across scales
- Determine spike time

The algorithm can be schematically seen on figure 3.6.

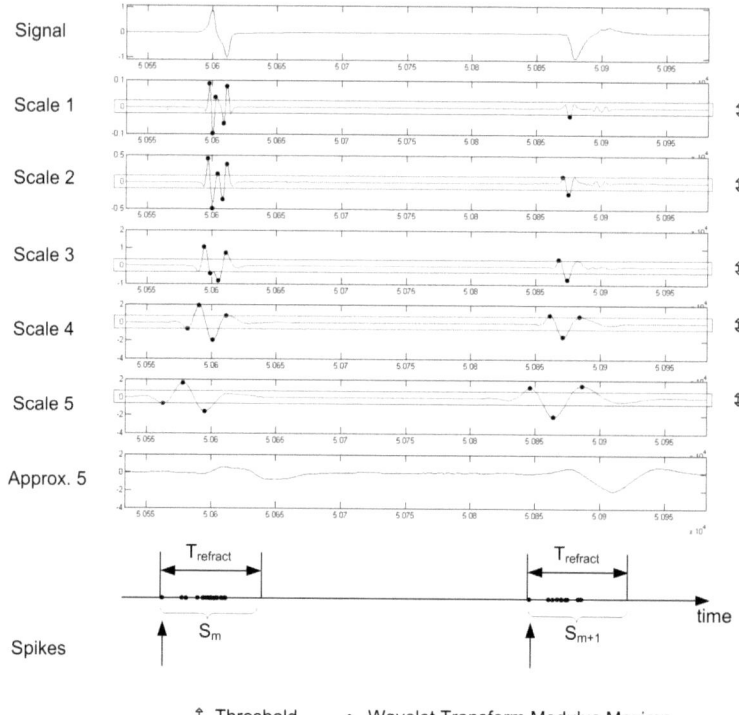

Figure 3.6 : Simple wavelet-based spike detection. $T_{refract}$ is the time window from the beginning to the end of the spike. Within this interval the algorithm does not detect any new spikes. A threshold selects the wavelet coefficients from each scale and merges them to a contiguous region S_m (bottom). The spike arrival time can be estimated in different ways.

The first two steps in the algorithm are equivalent to the first steps of denoising. The selection of the WTMM is new. In fact, Mallat [140] showed that the local maxima of the absolute value of the wavelet transform (WTMM) from high scales to low scales converge to the discontinuities of the signal. The concept of the detector is to determine those WTMMs. The next step in the algorithm is to process the WTMMs in order to find the spike arrival time. One way is to take the time of the maximum WTMM. Another method involves averaging of the locations of the WTMMs across scales [89]. Here, the first appearance of a WTMM after a given refraction time $T_{refract}$ will be attributed to a new spike

3.3 Spike Detection

event (see figure 3.6).

As mentioned in section 3.2 and in the appendix, there are several kinds of discrete wavelet transforms. They differ in shift-variance and computational complexity. For the sake of real-time implementation, the following four methods are compared in more detail:

- Thresholding in the time domain
- Orthogonal DWT
- Augmented DWT
- Shift-invariant DWT

The first method is widely used for simple and efficient detection in the time domain. Assuming Gaussian noise models, thresholds from $3\sigma_n$ to $5\sigma_n$ are commonly used. σ_n^2 corresponds to the noise power. The second method is Donoho's classical denoising method using the orthogonal DWT (figure 3.7a top). It is computationally highly efficient since its complexity is $O[N]$. The augmented DWT (figure 3.7a bottom) is similar to the orthogonal DWT except for the missing downsampling stage in the detail branch. Its complexity is also $O[N]$, however it requires more hardware resources than the DWT. The fourth method is the SIDWT. Its complexity amounts to $O[N \cdot \log(N)]$.

A reference signal consisting of three different neuronal waveforms was used (figure 3.7b). The simulated signal is based on a Poisson firing model, a refraction time $T_{refract}$ (i.e. recovering time of neuron) and a white Gaussian noise process. The spike templates were taken from real acquisitions on *in vitro* cortical cultures and from simulated models [44]. The comparison of the different detectors is based on (i) false positives (FP) and (ii) false negatives (FN) with respect to the simulated reference signal. FP is defined as the number of spikes

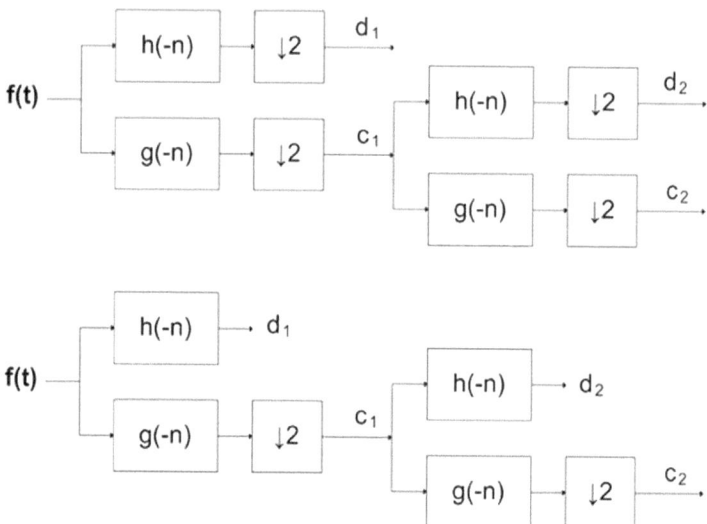

Figure 3.7 : Filter bank for orthogonal DWT (top) and augmented DWT (bottom) (a). Three spike templates are used for the generation of a reference spike train (b).

which were detected by the algorithm but that are not present in the reference train. FNs represent the number of spikes which were missed by the detectors. FP and FN are reported as a function of SNR in figure 3.8. The SNR is defined as the ratio between the maximum spike amplitude (normalized to 1) and three times the standard deviation of the noise (see equation 3.5).

The graphs show that for low SNR the SIDWT-based detector performs better with respect to FPs and FNs than the conventional threshold based spike detector. The shift-invariance and the compaction property of the wavelet transform enables a good overall performance. The fast DWTs (i.e. orthogonal and augmented) do hardly detect any false spikes (i.e. low FP), however they also miss many events that were present in the reference train (i.e. high FN). The shift-variance compromises the detection performance with respect to the shift-invariant DWT. In general, the thresholds for the wavelet coefficients seem to

3.3 Spike Detection

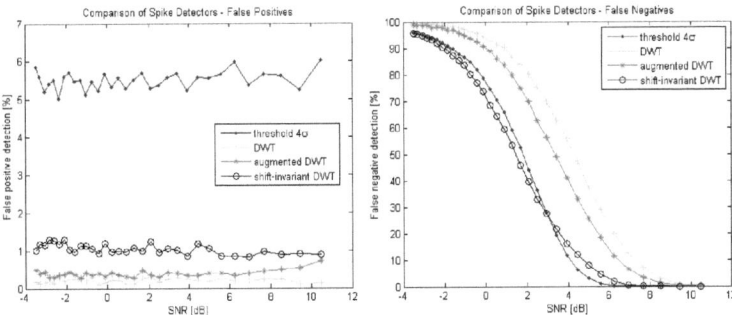

Figure 3.8 : Performance of spike detectors: based on time domain threshold, on orthogonal DWT, on augmented DWT and on shift-invariant DWT. The false positive curves (a) and the false negative curves (b) are shown.

be too conservative and they can be optimized with respect to the given nature of the spikes. The FP is lower for all wavelet-based detectors since the combination of several criteria at different scales reduces the risk to detect false spikes.

The wavelet transform can also be interpreted as a nearly matched filter or correlator for a large class of signals [141, 142]. Intuitively, one expects the wavelet transform to perform better in high noise environments than simple time domain based threshold detectors. In fact, the SIDWT outperforms the conventional threshold method for very high noise levels, as one can see in figure 3.8, but at the cost of higher computational complexity. Hence, the DWT-based detectors are not necessarily better than the threshold detector (see FN graph in figure 3.8b). The real advantage of the DWTs appears when including spike sorting. This is the topic of discussion in the following section.

3.4 Spike Sorting

Wavelet-based sorting and classification methods using both redundant and non-redundant DWT systems were already proposed [87,92,137,138]. The non-redundant DWT uses orthogonal basis signals [143] and is particularly adequate for data compression. However, it is less appropriate for feature extraction because it is shift-variant. As can be seen in the appendix shift-invariance can be restored with overcomplete or redundant representations, e.g. non-decimated discrete wavelet transforms [143, 144]. Several authors [92, 137, 138] presented spike sorting methods based on non-redundant orthogonal DWT or on wavelet packets [87] avoiding the shift-variance problem by centering all detected spikes within an analysis window. However, for highly integrated MEAs, this approach is not practical for real-time implementation because of its prohibitive memory requirements for the storage of the signal samples of each channel in a window around the spike.

In general, it is essential to extract characteristic features from the spike form in order to classify them. A widely used technique is classification based on the first three principal components of a PCA. The approach we chose here is based on an alternative representation of the DWT. The energy-zero-crossing method (EZC) was proposed by Mallat [145] and presented as an adaptive sampling of the DWT in order to reduce shift-variance by reconstructing a signal based on its zero-crossings. The idea is graphically shown in figure 3.9. The DWT approximates the underlying continuous wavelet transform (CWT) of a signal. This is similar to the FFT that approximates the Fourier transform. A detailed discussion of the relationship between a DWT and the CWT can be found in [146]. The EZC method determines first the zero-crossings of the DWT. Subsequently, the *signed* energy between two zero-crossings is computed based

3.4 Spike Sorting

on the samples of the DWT and then normalized by the distance between the two zero-crossings. The resulting linear piecewise function (figure 3.9, shaded area) is more robust and near shift-invariant. It is thereafter used to compute a classification metric $f_{j,k}$ at each scale j and for each spike m.

A critically sampled DWT, such as the orthogonal DWT, gives an EZC representation that is still shift-variant (figure 3.9b). However, the oversampled (i.e. augmented) DWT leads to an EZC representation that is more stable with respect to translations (figure 3.9c, d) and remains computationally very efficient.

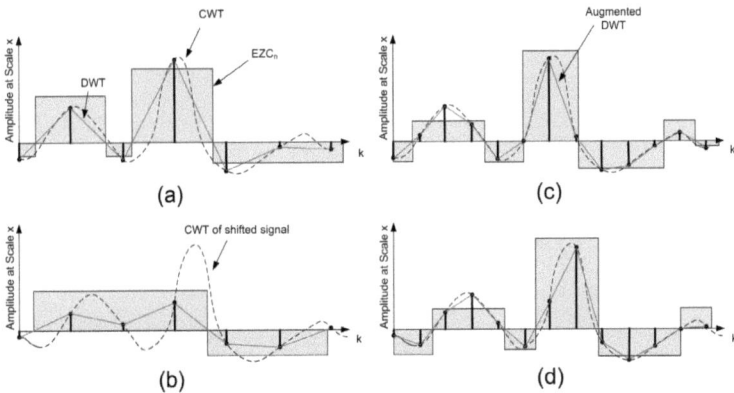

Figure 3.9 : The underlying continuous wavelet transform (CWT) of a signal is shown with the dashed line. The DWT approximates the CWT, however it is not shift-invariant. The energy-zero-crossings (EZC) is an alternative sampled representation of the DWT (a). The EZC is still shift-variant (b). An EZC of the augmented DWT is shown in (c). The EZC of the shifted DWT is less shift-variant (d).

The spike sorting algorithm based on EZCs involves following steps:

- Compute DWT and keep the detail coefficients d_k

- Locate spike m (see section 3.3)

- Determine K_j zero-crossings in a window W around spike m and for all scales j $\rightarrow ZC_{j,k}$ for $k \in [1, K_k]$

- Compute energy $E_{j,k}$ between zero-crossings $ZC_{j,k}$ and $ZC_{j,k+1}$
- Normalize $EZC_{j,k} = \frac{E_{j,k}}{ZC_{j,k+1}-ZC_{j,k}}$
- Compute metric $f_i(m)$ for spike m at scale j using all $EZC_{j,k}$ of that spike
- Cluster $f_j(m)$ for the most important scales

We defined the following metric f_j:

$$v_j(m) = a + i \cdot b \begin{cases} a = \sum_k EZC_{j,k} & \text{for } \forall \text{ k with } EZC_{j,k} \geq 0 \\ b = \sum_k EZC_{j,k} & \text{for } \forall \text{ k with } EZC_{j,k} < 0 \end{cases} \quad (3.8)$$

$$f_j(m) = |v_j(m)| \quad (3.9)$$

A block diagram of the real-time spike sorting algorithm is shown in figure 3.10. The wavelet coefficients consisting of a set of augmented detail coefficients $\{d_j\}$ and a set of approximation coefficients $\{c_j\}$ are obtained at each significant scale. In a next step, the spikes are detected using a thresholding process on the DWT. The threshold is computed according to Donoho's estimator during an initialization phase. Spike m is detected when there is a contiguous region $S(m)$ of coefficients, merged from all scales, that exceed the threshold (see section 3.3). The wavelet coefficients within each region $S(m)$ are used to estimate the zero-crossings of the wavelet transform. Furthermore, the energy between two subsequent zero-crossings needs to be estimated. The $EZCs$ from the significant scales within $S(m)$ are used to build a discriminant measure $f_j(m)$ for spike m and scale j, such that it can be employed for spike clustering. A simple measure f_j is the absolute value of the complex number consisting of the sum of all positive $EZCs$ within $S(m)$ for the real part and of the sum of all

3.4 Spike Sorting

negative $EZCs$ within $S(m)$ for the imaginary part (equations 3.8 and 3.9). It can be seen from figure 3.10 that only a small set of simple operations (i.e. additions, multiplications and thresholding) are required to perform all necessary computations involved in successful neuronal spike detection and sorting.

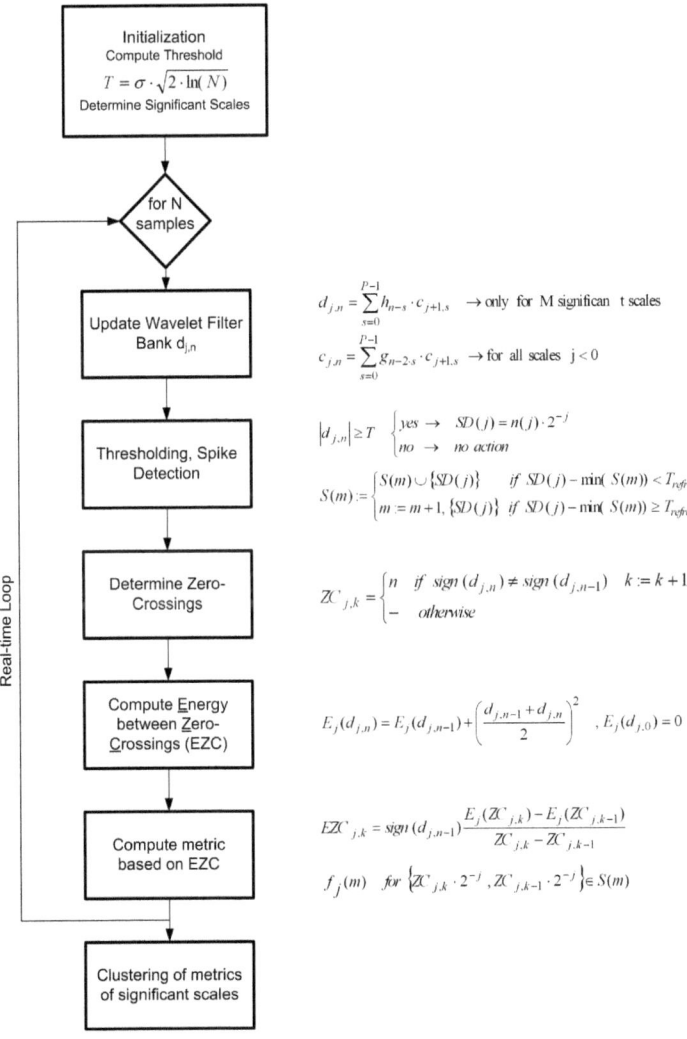

Figure 3.10 : Real-time spike sorting algorithm

The issue of shift-variance for feature extraction, and thus for spike sorting and classification, is illustrated in figure 3.11. In a series of runs an arbitrarily delayed spike of one type from figure 3.7b in a time window of 50 ms was generated. For each run an orthogonal 8-scale DWT was performed and a feature measure f_j based on EZC was computed using the detail coefficients d_j at each scale. The significant scales correspond to the time resolutions at which the signal characteristics were expected to appear. The discrimination measures of the three most significant scales (scales 5, 6 and 7 in this example) were plotted in a 3d-graph for 150 runs. It can be observed that the three different types of spikes do not generate well separated clusters for the non-redundant DWT and therefore efficient spike sorting becomes difficult (figure 3.11a). The shift-variance of the DWT produces significant overlapping of the three classes of spikes. Figure 3.11b shows the clustering of the same signals based on an augmented (redundant) DWT where subsampling by 2 of the detailed coefficient after each level decomposition was omitted (see appendix). The redundant DWT clearly shows an improved clustering.

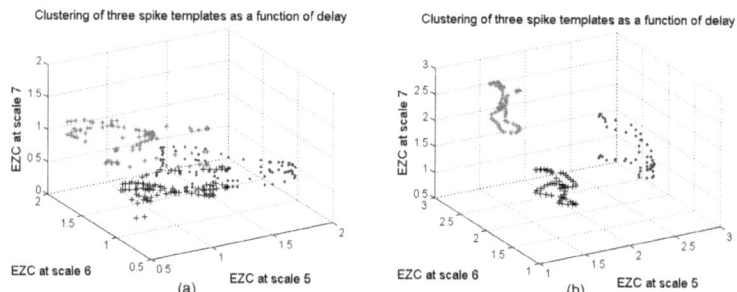

Figure 3.11 : 3d-plots of the discrimination measures of the three most significant scales (a,b). The measures are based on EZCs and computed by the orthogonal non-redundant DWT (a) and by the redundant DWT (b).

Figure 3.12a and b show the clustering effect of the wavelet-based algorithm for a model spike train containing the three different spike forms at different SNRs

3.4 Spike Sorting

(9×type 1, 16×type 2 and 4×type 3). For illustrative purposes the standard PCA-based method (described in figure 3.1) was also applied to the signal and the resulting clustering can be compared on the same figure.

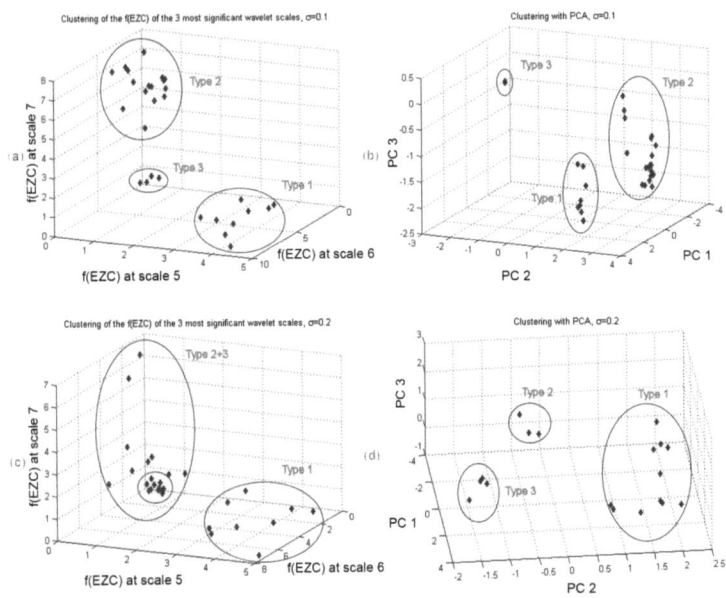

Figure 3.12 : Clustering for spike sorting under different SNRs. The wavelet-based algorithm separates the spike forms for a SNR of 4.4 dB (a). The same signals with the same SNR are applied to the PCA-based sorting algorithm (b). The wavelet-based algorithm cannot discriminate spikes of type 2 and 3 anymore for a SNR of -1.6 dB, however it still detects all spikes (c). For the same SNR the PCA-based method still separates the spikes, however the simple threshold spike detector does not detect all spikes (d).

Two cases for SNRs of 4.4 dB and -1.6 dB, respectively, are shown in figure 3.12. The PCA-based method discriminates three families for the spikes that were detected (figure 3.12d). The first three principal components carry the most relevant characteristics of the three spike types and thus they produce well separated clusters for the waveforms of figure 3.7b and d. However, a lot of events do not appear in the cluster plot because the low SNR compromises the spike detector in the time domain.

For the lower SNR in this example, the clusters for wave types 2 and 3 are not separable anymore with the wavelet-based method. This is due to both a low SNR and a not yet optimal discriminant measure $f_j(m)$ that has moderate clustering properties for similar waveforms. However in general, it can be seen that the wavelet-based method enables good clustering that is essential for the sorting of different spike families. This demonstrates that simple wavelet-based sorting, as proposed here for real-time implementation, is a good alternative to the computationally expensive PCA-method.

Finally, the algorithm was applied on recorded data from the APS-MEA. The output signals from the high-density APS-MEA suffer from drifts, coupling and noise. A calibration circuit was included to compensate for drifts (see chapter 2.2.2). Denoising is handled by methods presented in section 3.2. However, digital coupling and compensation also introduce artefacts on the channels. Examples of the signal types that were observed on APS-MEA recordings are shown in figure 3.13a. Type 1 corresponds to a transient due to a calibration step and type 3 consists of a signal that was observed on electrodes which were not optimally operational (i.e. due to fabrication tolerance and interface effects). Type 2 is a typical spike recorded from a cortical culture. Figure 3.13b shows the cluster plot that was applied to arbitrarily shifted versions of the signals from figure 3.13a. The clustering was achieved with the fast augmented DWT.

3.5 Conclusion

In this chapter we discussed the different signal processing steps that are necessary to extract the important signals from the raw recordings of the APS-MEA system. We showed that all steps, such as denoising, spike detection and spike

3.5 Conclusion

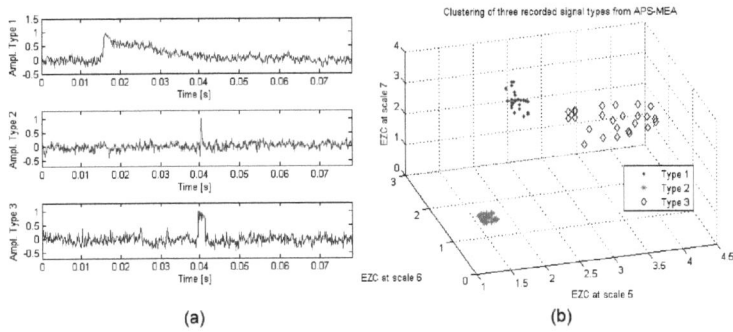

Figure 3.13 : Three recorded signal types from the APS-MEA (a). Spike clustering of the three types using the fast augmented DWT (b).

sorting, can be achieved with a unified wavelet-based framework. This framework is particularly well suited to implementation in hardware. We believe that signal processing in hardware is a critical step towards an efficient data handling and signal analysis of recordings from high-density MEA systems.

Since the high-density MEA platform is limited in terms of integration area and power per electrode, the system is intrinsically noisier than conventional low-density MEA systems. The denoising performance of the shift-invariant discrete wavelet transform is better than the one of the fast (orthogonal) transform, however at the cost of increasing the computational complexity from $O(N)$ to $O(N \cdot \log N)$. It is therefore appropriate to use the fast DWT for real-time or on-line denoising implementation and the SIDWT for off-line processing.

Spike detection can also be done in the wavelet domain. A critical drawback of the currently implemented simple spike detector is the lack of discrimination of overlapping spikes or closely firing neurons. The user-defined refractory period $T_{refract}$ in our algorithm disables any capacity to detect spikes that are closer than this minimum value. In general, the SIDWT turned out to be more efficient than the conventional time domain thresholding for very low SNR signals. The fast orthogonal DWT using Donoho's universal threshold is more conservative than the SIDWT or time threshold method. The wavelet coefficient threshold can be optimized with respect to the nature of the signal, however a conventional time domain thresholding seems still a good alternative for simple open-loop operation of the system.

A major advantage of the wavelet-based signal analysis besides denoising is its spike sorting capacity. The fast DWT is not optimal for sorting due to its shift-variance. Hence, we introduced a simple adaptive resampling of the wavelet transform using EZC as well as an augmented fast DWT to enable a near shift-invariant method for wavelet based spike sorting. The algorithm turned out to be efficient for both spike sorting and discrimination of unwanted interferers. However, the discriminant measure f_j is very critical to identify individual clusters. Here, we introduced a simple operator (equations 3.8 and 3.9) that, in turn, clearly limits the discrimination for waveforms that have similar absolute energy distributions at the same scales.

Contributions

To the best of our knowledge, the following contributions can be considered as original:

- Elaboration of a wavelet-based framework for denoising, spike detection and spike sorting of *in vitro* neurophysiological signals

- Elaboration, implementation and comparison of DWT-based real-time spike detectors

- Proposition and investigation of a fast feature extraction algorithm for spike sorting using energy-zero-crossings and an augmented DWT

Further potential research

- Computational efficiency of the wavelet transform is based on Mallat's algorithm [147]. Although it is already very fast, i.e. with a complexity of $O(N)$, significant computational savings can be achieved using FFT-based algorithms [148] or fast running FIR algorithms [149] for the computation of wavelet coefficients in a real-time environment

- Optimization of denoising, spike detection and classification using adaptive wavelet bases and wavelet packets [150]

- Investigation of complex dual-tree wavelets for shift-invariance [151]

Chapter 4

Image-Based Signal Processing

In the previous chapter we showed how the discrete wavelet transform can be used to enable real-time processing of large-scale and high-density MEA recordings. The methodology was based on a channel-by-channel analysis, i.e. each electrode was processed independently of its neighbourhood. With electrodes separations in the order of several hundred of micrometers, MEAs are unlikely to measure same neurons on different recording sites. With the downscaling of electrode separations signal contributions from individual neurons or local group of neurons are likely to occur on multiple adjacent electrodes. In particular, locally generated field potentials can be captured by several electrodes. Also, a single neuron might be coupled to several adjacent electrodes. Thus, signal sources can be measured on different electrodes. As a consequence, the recorded channels are spatially correlated and contain a certain degree of redundancy. Spatial correlation between multiple electrodes has already been shown and exploited in *in vivo* experiments using tetrodes and microelectrodes [152], however it has not been investigated in *in vitro* neuronal cultures mainly due to the lack of high-resolution devices.

Correlation in neuronal networks has to be clearly defined since two different types of correlation are encountered in multielectrode recordings: (i) intrinsic neuronal correlation (dependence) between firing patterns of neurons with an underlying physiological meaning [15, 96] and (ii) signal correlation where one or several signal sources are measured on multiple recording sites. Neuronal correlation is subject to detailed investigation, ranging from research activities on cultured retinas [37] to *in vivo* studies of neuronal interactions [15]. Signal correlation is a general method of array signal processing and statistical signal processing and enables applications in fields, such as radars, seismology and communications [78].

In array signal processing multiple sensors capture the signal of waves propagating in space [153]. The fusion of spatio-temporal information from different locations (i.e. array of sensors) enables to enhance the detection of the propagating signal and allows tracking of involved sources. It is also referred to as spatial filtering or beamforming since in its original form linear filter theory was applied to a set of sensors [154] to improve the sensitivity of the system with respect to a precise spatial direction. A review of array signal processing can be found in [78]. Rao introduced wavelets for nonstationary signals in multisensory signal estimation [72]. Discrimination and classification of neural activity using multisensory recordings has been explored for many years [71, 155, 156]. An array denoising technique applied in the wavelet domain for neuroelectric data from *in vivo* electrodes was proposed by Oweiss [73]. In general, all those techniques can be applied to our high-density MEAs, however these algorithms are computationally very expensive and hence, they are not very practical for real-time/on-line use in our system.

The ultimate high-resolution of the APS-MEA system with respect to the electrophysiological activity of a neuronal network naturally led to an imaged-based data acquisition and representation where each electrode is interpreted as a pixel of an image. It is therefore obvious to look for interesting concepts in the image/video field that could be applied for processing and analysis of neuroelectric data. In particular, signal enhancement techniques can potentially benefit from image-based methods. For instance, feature- and content based interpretations of images allow increasing SNRs [157]. Moreover, alternative biological characterisation of networks might be enabled through image/video processing as well.

In that sense, wavelets are widely used in image processing for (i) denoising and compression [139, 158], (ii) edge and contour detection [140], (iii) feature extraction [159] and texture characterisation [160]. Spatial correlation of "real-world" images is expressed by correlation of adjacent scales in the wavelet transform [161]. The inherent multiresolution concept of wavelets [143] can even potentially enable a cellular-to-network characterisation of electrophysiological activity.

The objective of this chapter is to explore potential new concepts that arise from the high spatial information of the large-scale, high-resolution MEA acquisition system. The discussion is limited to a signal processing point of view. Any relevance of high-resolution information with respect to underlying biological processes is not addressed in this work. Section 1 briefly describes the signal model that was used. Section 2 addressed redundancy and spatial correlation from biological signals measured with the high-resolution MEA system. Section 3 discusses denoising including additional spatial information and section 4 gives an outlook on a potential multiresolution characterisation of biological data acquired at a large scale and with high resolution.

4.1 Signal Model

Models of physical phenomena are fundamental to a qualitative and quantitative understanding of the involved underlying processes. They are also critical for evaluating the extraction of important system parameters or characteristics by computer based tools. In the one-dimensional signal processing case, as discussed in chapter 3, it was appropriate to use state-of-the-art firing models [109] and signals [44] for the evaluation of the channel-based algorithms. However, these models need to be appropriately coupled to a large-scale high-resolution array. Hence, signal processing and high-resolution monitoring of large neuronal networks with thousands of cells has also to be assessed with an appropriate reference data model.

NEURON [162] is a powerful simulator that is well adapted to small network of neurons. However, it uses a detailed Hodgkin-Huxley cell model and therefore, it is not meant to efficiently simulate large neuronal networks coupled to high-resolution MEAs. In this regard, André Garenne, at the INSERM Bordeaux, developed a simulator for large-scale, high-resolution MEA signals [163].

The simulator is based on a cell model from Izhikevich [164–166] and approximates the Hodgkin-Huxley model with only a few parameters.

$$\frac{dv}{dt} = 0.04v^2 + 5v + 140 - u + I \tag{4.1}$$

$$\frac{du}{dt} = a(bv - u) \tag{4.2}$$

$$\text{if } v \geq 30\,\text{mV then } \begin{cases} v \leftarrow c \\ u \leftarrow u + d \end{cases} \tag{4.3}$$

v represents the membrane potential of the neuron and u represents a membrane

4.1 Signal Model

recovery variable with a negative feedback to v. u also accounts for the activation and inactivation of K^+- and Na^+-channels, respectively. I corresponds to synaptic or DC-injected currents. a, b, c and d are equation parameters that allow to implement many different cell types connected by axons and synapses (Izhikevich, 2003). The synapses are simulated with a standard conductance model [29, 109, 167].

An additional parameter introduced in the modelling is the cell-to-electrode coupling. The extracellular current from a cell and recorded by an electrode was approximated by the first derivative of the cell's membrane potential [168]. Furthermore, the total incident current on one electrode is modelled as the sum of the currents of all cells coupled to a single electrode. Finally, the neuronal network is statistically built by using following (non-exhaustive) list of parameters:

- Cell types, i.e. parameters a, b, c and d of the Izhikevich model for different neuron types [164] average cell body radius.

- Distribution of cells, i.e. deterministic or statistical geometry (e.g. grid, sphere, etc.)

- Synapse type, i.e. includes characteristics such as plasticity rules, axonal conductance speed, synapse delay, synaptic weight, reverse potential, connectivity, etc.

This simulator that is named INNS (Izhikevich Neural Network Simulator) was kindly provided by André Garenne, INSERM. Quantitative evaluations of specific image-based algorithms in this chapter were performed by a reference data set that was obtained from the INNS.

4.2 Redundancy of APS-MEA Signals

Do high-resolution MEA signals carry redundant information that can be used to increase the overall signal quality and the discrimination of overlapping sources (i.e. neurons)?

With electrode sizes of $21\,\mu m \times 21\,\mu m$ and electrode separations of $21\,\mu m$, the high-resolution MEA platform converges to the dimensions of neuron somas. Figure 4.1 illustrates the scale of the neurons and the recording electrodes. It seems obvious that in such a configuration a neuron can contribute signal components to several electrodes and, vice versa, individual electrodes can record from several neurons. Inherently, one would expect a substantial amount of redundant information between different adjacent electrodes.

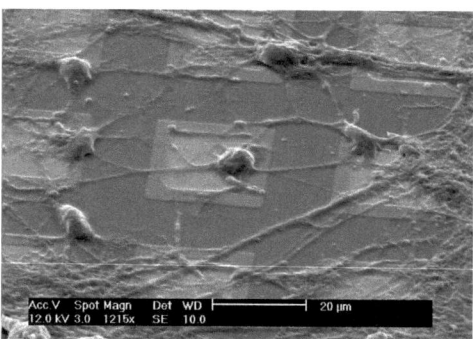

Figure 4.1 : Scanning electron microscope (SEM) photograph of a dissociated neuronal cell culture with underlying dummy electrodes featuring the size of the APS-MEA. Courtesy of Luca Berdondini, Samlab.

In order to answer above question one needs to refer to multivariate signal processing. Multivariate statistics and array signal processing extract additional information from multiple recording sites in order to improve the signal quality and/or to identify individual signal sources. Concepts that were initially used in speech processing and communications are now widely employed in computational neuroscience, such as for enhancement of signals from electroen-

4.2 Redundancy of APS-MEA Signals

cephalography (EEG) or for identification of independent brain regions in functional magnetic resonance imaging (fMRI). Mathematical techniques include PCA and independent component analysis (ICA) [77]. The methods were both applied to simulated [79] and *in vivo* [36] multielectrode recordings to separate different neurons from overlapping action potentials [169]. PCA removes co-variance between multiple channels. ICA minimizes mutual information in the signals. Since independence is a stronger statistical property than co-variance or correlation, true redundancy reduction is achieved by ICA whereas PCA removes it only partially.

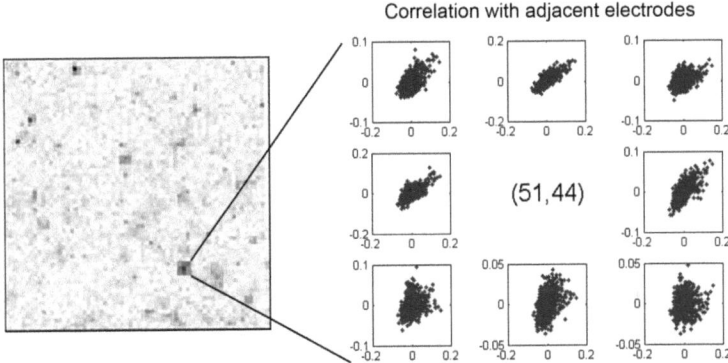

Figure 4.2 : Scatter plot of the eight adjacent electrodes of electrode (51,44) (experiment with dissociated cortical cultures from rat). The different degrees of correlation can be seen. An uncorrelated electrode pair converges to a rectangular cloud, i.e. the electrode (50,45) (top right), (52,43) (bottom left) and (52,45) (bottom right) are only little correlated with the center electrode. A high correlation is expressed by a stretched elliptical shape of the cloud. The correlation is shown for a time window of 130 ms during the occurrence of a burst. The x-axis in each plot relates to the central electrode, the y-axis is the corresponding adjacent electrode.

A qualitatively simple way to demonstrate that measurements performed with the high-resolution MEA system carry redundant information is to plot the signals from one electrode with respect to its neighbours. Figure 4.2 shows a typical example of an acquisition from a cortical culture performed at INSERM, Bordeaux. Correlations between the x- and the y-axis can be clearly seen since most clusters have specific directions of minimum and maximum variance. Such spatial redundancy is necessary if one wants to perform spatial denoising.

4.3 Spatio-Temporal Denoising

4.3.1 Context

In recent years many methods for image denoising were developed. An exhaustive review can be found in [170]. Most algorithms were designed to perform well for specific type of images, i.e. "real-world" images or medical images [142]. However the type of images from high-density MEAs is new and therefore, it has never been subject to an investigation with respect to potentially applicable algorithms. Here, I will focus on basic denoising methods involving the wavelet transform. The wavelet framework has been adopted in this work for several reasons, i.e. ease of real-time implementation and as a "general purpose" tool for signal enhancement, spike detection and spike classification for high-density MEA recordings (see chapter 3).

Section 4.2 outlines that spatial redundancy exists in the recordings of the high-density MEA. However, it is not clear yet how this redundant information can be used to increase the signal quality. In the previous section it was also mentioned that PCA/ICA reduces redundancy in multivariate signals. Redundancy reduction is equivalent to decorrelation of a signal. Moreover, decorrelation is also referred to as whitening [77] and reduces the autocorrelation of a signal, or equivalently, maps the resulting signal closer to white noise. In that sense, any transform that reduces the autocorrelation of the transformed signal can be used for redundancy reduction (i.e. Karhunen-Loève Transform, Discrete Cosine Transform, etc.). In particular, the wavelet transform also decorrelates signals, hence a spatial (2d) wavelet transform on the data from APS-MEA recordings is expected to perform a decorrelation in space. When denoising in the spatial domain first and followed by one in the temporal dimension, a denoised signal of higher quality compared to a temporal denoising of each electrode independently should be obtained.

4.3.2 Comparison of Multidimensional Denoising

Multidimensional wavelet-based denoising was originally developed for images [171]. Many algorithms were proposed in order to take into account the intrinsic spatial correlation of natural images [157, 161, 172]. Recently, wavelet domain denoising was also introduced in video technology [173–175] and 3d-imaging [174, 176]. In most of the cases the wavelet threshold selection is crucial for the performance of the wavelet-based denoising. Hence, several thresholding methods were proposed [135, 177], i.e. spatially adaptive thresholds [139], in order to improve the denoising of images.

Three simple methods will be compared in the following (figure 4.3):

- DWT : Orthogonal 1d-DWT denoising, channel-by-channel as discussed in section 3.2

- PCA-DWT: A hybrid PCA-Orthogonal DWT-based denoising, PCA spatial decorrelation followed by a channel-by-channel time denoising

- (2d+t)-DWT: Multidimensional WT-based denoising, each frame with 2d-SIDWT and then the resulting frame sequence in time

Method 1 corresponds to the conventional 1d-denoising that was discussed in section 3.2 and serves as reference method. Method 2 decorrelates each individual frame using the PCA method [77]. The new frame sequence is then denoised with the 1d-denoising as in method 1. Finally, method 3 is based on two independent wavelet transforms, where denoising takes place in the spatial domain first and then in the time domain. For the spatial wavelet transform a shift-invariant DWT (SIDWT) similar to the one described in section 3.2 was preferred. Otherwise, the orthogonality of the DWT would significantly introduce block artefacts when reconstructing the data from the thresholded coefficients.

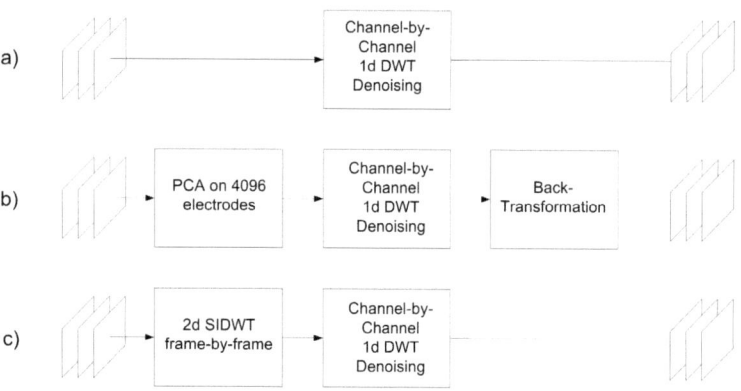

Figure 4.3 : Orthogonal 1d DWT- (a) , PCA-DWT- (b), and (2d+t)-DWT-denoising (c).

For a first test sequence to compare the three methods one needs simple geometric objects with high spatial correlation. Therefore, a frame sequence (i.e. 1024 frames) with two moving rectangles was chosen (figure 4.4b). We assumed a white Gaussian noise process with different variances to alter the input SNR. The output SNR was then computed as follows:

$$SNR_{out} = \frac{\sum_{t=0}^{T} \sum_{x=0}^{M-1} \sum_{y=0}^{N-1} \hat{f}(x,y,t)^2}{\sum_{t=0}^{T} \sum_{x=0}^{M-1} \sum_{y=0}^{N-1} \left[\hat{f}(x,y,t) - f(x,y,t)\right]^2} \quad (4.4)$$

where the size of a frame is M x N, T is the number of frames in the sequence,

4.3 Spatio-Temporal Denoising

$\hat{f}(x, y, t)$ are the denoised output frames and $f(x, y, t)$ are the original frames.

The output of the 2d-wavelet transform consists in three detail components, i.e. the horizontal H, vertical V and diagonal D (see appendix) details. Therefore, three different thresholds were defined from the finest scale of each detail. The wavelet coefficient threshold was fixed to $3 \cdot \sigma_{n,H/V/D}$, where $\sigma_{n,H,V,D}$ is similar to the one-dimensional case (equation 3.1) for each finest detail H_1, V_1 and D_1. $3 \cdot \sigma_{n,H/v/D}$ was chosen in order to be less conservative than Donoho's universal threshold, which overly smoothes images for high number of samples, i.e. image size M x N (64x64).

In all experiments the Haar wavelet was used for the spatial WT and Symlet3 was used for the temporal WT.

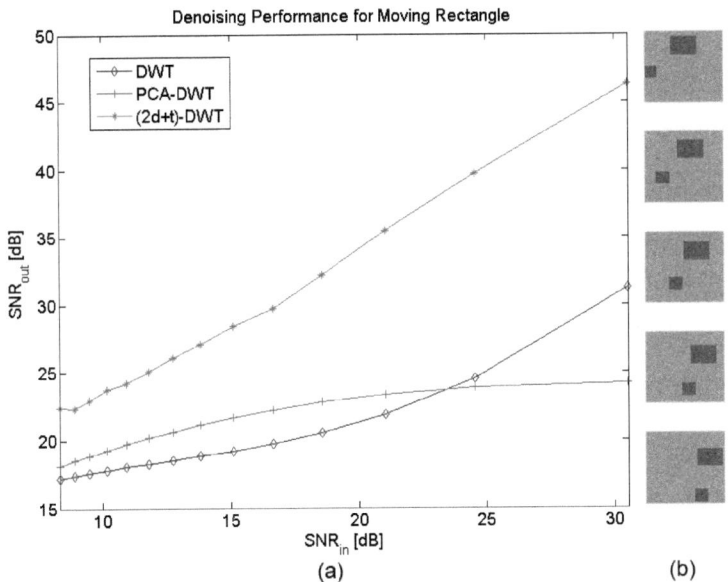

Figure 4.4: Denoising performances of the three methods for a synthetic 'Moving Rectangle' frame sequence (a). Excerpt of five equidistant frames with moving objects (b).

The results from figure 4.4 show that the (2d+t)-DWT performs better than the two other methods. The degradation of the PCA-DWT method with respect to the DWT method at high SNRs is due to the fact that spatial correlation inherently exists in each frame. The temporal change (i.e. motion) includes only little information about the spatial relationship of the objects. Therefore, decorrelation that is based on temporal relationships between channels, as in the PCA case, is not adequate for moving well defined geometrical objects. How are the three methods performing for modelled neurophysiological signals?

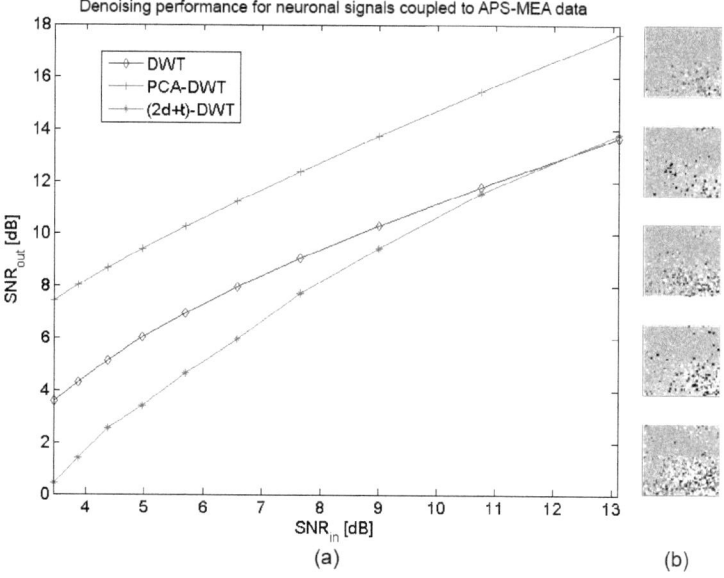

Figure 4.5 : Denoising performances of the three methods for a simulated burst sequence coupled to an APS-MEA. The INNS was used to generate the reference data.

One can see from the results shown in figure 4.5a that the PCA-DWT method gives the best denoising performances of the three tested methods for simulated APS-MEA data. For this type of signals the spatial decorrelation of method 3 degrades the overall denoising performance with respect to the other two methods. This is due to the specific type of signals. By looking at figure 4.5b we notice that the simulated burst on the APS-MEA appears locally very much like noise when each frame is considered separately (or independently) of the other frames. It is obvious that noise like frames, i.e. without any clear structure, can not be further spatially decorrelated. Contrary to the PCA case, 2d-wavelet-based spatial denoising relies on the signal statistics from only one frame, without using the temporal dimension, whereas the PCA-based method takes into account time information to decorrelate in space. This implies that for simulated neurophysiological bursts one can not really benefit from the additional high spatial resolution by only considering the two spatial dimensions independently of the time domain. The denoising performance for measured neurophysiological signals is expected to be somewhere between the case of moving rectangles and the one of large network burst. It will be necessary to look at the more general *non-separable* 3d-wavelets [174] in order to enable multidimensional decorrelation.

4.3.3 Discussion

1d-DWT does not have as good performance as the PCA-DWT method for neurophysiological signals from high-resolution MEAs. However, due to the significantly lower computational complexity, it remains the method of choice for on-line or real-time processing, whereas the PCA-DWT method is rather adapted to offline processing. It has been demonstrated that high-resolution of electrodes does not lead to an improvement in signal quality when using the decorrelation property of the fast discrete wavelet transform applied to the spatial and temporal dimensions independently. Simulated bursts obtained by the INNS do not show any significant spatial correlation at a single frame level as opposite to the moving rectangle example with its geometrically well defined objects.

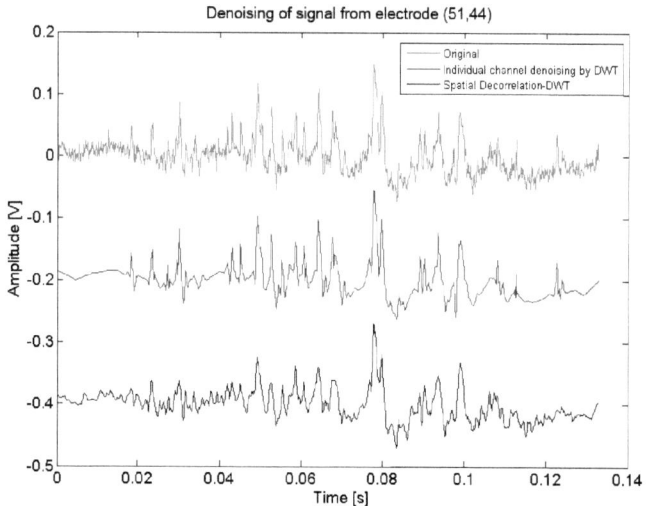

Figure 4.6 : Recorded burst from a cortical culture (INSERM, DIV23) (top). DWT-denoised signal where each channel was processed separately (middle). Spatial decorrelation by PCA, followed by a channel-by-channel DWT denoising (bottom).

In that sense, *separable* 3d-filter banks (i.e. 2d + time) do not lead to an improvement of the SNR of high-resolution MEA signals. Spatio-temporally correlated patterns will have to be searched for with *non-separable* wavelet kernels [178].

Figure 4.6 gives an example of signals obtained by PCA-DWT filtering and by DWT filtering only.

4.4 Multiscale Analysis - Global vs. Local Activity

The presented high-density MEA acquisition system with 4096 electrodes on $7\,mm^2$ leads to a high-resolution activity representation of a large network of neurons. Conventional low-resolution MEAs considers the activity as a set of point processes and the characterisation of the network dynamics is mainly done

4.4 Multiscale Analysis - Global vs. Local Activity

by extracting statistical information between pairs of electrodes (i.e. cross-correlogramms, mutual information). Characterisation of networks such as burst propagations [11, 102, 179–181] and activity sources [182] can only be done at a very coarse spatial level. Also, the dynamics of the network are difficult to catch in on-line or real-time modes. For instance, burst statistics are computed over a set of burst occurrences and are therefore only available after a long period of experimentation.

The high-resolution performance of the MEA system allows recording the details of activity propagations down to the cellular scale. On the other hand, a large neuronal network can be monitored with the platform and it might not be necessary to know the details at the cellular level. Both high-resolution and large-scale features of the system enable analysis at different level of details (i.e. zoom mode). The wavelet transform inherently provides an appropriate tool to perform analysis at different resolutions. By applying each frame to a 2d-wavelet filter bank, the signals at each scale give the amount of relevant information at the corresponding level of detail. Fine spatial variations in the activity will be enhanced at lower scale, whereas slow variations across the array will be pointed out at higher scales. Such a spatial decomposition can be performed by a 2d-filter bank that splits the input signal into four bands at each level: (i) approximation, (ii) horizontal detail, (iii) vertical detail and (iv) diagonal detail. Thereafter, the approximation can be fed to the filter bank of the next level in order to further decompose the frame at lower resolutions. The concept is equivalent to the one presented for one-dimensional signals in chapter 3.

An illustrative example is shown in figure 4.7. The original 2d-signal is displayed (left) together with its first five approximations of a shift-invariant DWT (i.e. Haar wavelet). Each scale can be considered as an average showing details at a given resolution.

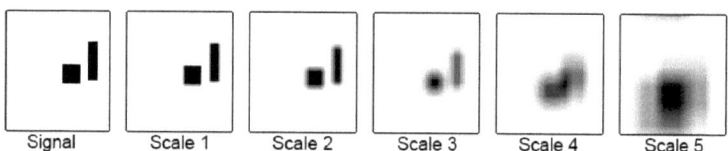

Figure 4.7 : The original 2d-signal is shown on the left. The five following frames show, from left to right, decreasing levels of details. For instance, scale 4 and 5 provide a sort of "frame" averages of the two original objects.

Applying this idea to high-density recordings one can obtain a multiresolution representation [147] of a parameter of interest. Depending on the interest of details one can choose the corresponding scale. Obviously, all information from a higher scale can already be obtained from a lower scale, however depending on the application one might only be interested in a coarse quantification of the network and further details only compromise the visualization. This spatial quantification can also be considered as a sort of information compression where only the data relevant to an experiment is retained. For instance, higher scales are typically transmitted first in communication links [183]. At the receiver end the image is reconstructed starting from higher scales and therefore quickly provides a coarse view while refining the image little by little.

An illustrative view of such a multresolution representation is illustrated in figure 4.8. Here, a simulated burst (see section 4.1) with is spatial wavelet decomposition is depicted for different time instants.

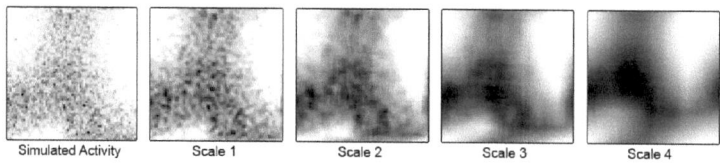

Figure 4.8 : Simulated burst activity of a neuronal network on the APS-MEA (left). Activity is shown in black. The four other frames correspond to the approximation of the wavelet decomposition at each level.

4.4 Multiscale Analysis - Global vs. Local Activity

Chao introduced a global activity measure, the center of activity (CA) for the visualization of activity and plasticity of dissociated cortical cultures on conventional MEAs [184]. In an analogous manner, one can define a center of gravity (CoG) on the acquired high-resolution frames in order to characterize the instantaneous state of the activity of a neuronal network. The trajectory of such a metric with respect to time allows quantifying the network dynamics at a required level of detail. The CoG is defined as follows:

$$CoG_j(\vec{r_0}) = \frac{\sum_{i=1}^{N} \vec{r_i} \cdot V_j(\vec{r_i})}{\sum_{i=1}^{N} V_j(\vec{r_i})} \qquad (4.5)$$

Where $\vec{r_0}$ corresponds to the location of the CoG, $\vec{r_i}$ is the vector to the pixel i and $V_j(\vec{r_i})$ is the activity level of the pixel i at scale j. The function $V_j(\vec{r_i})$ can be defined as any function that measures the instantaneous activity at a pixel of scale j. A non exhaustive list of measures is:

- raw signal from electrode
- instantaneous spike rate
- windowed (moving) power/variance of the signal
- windowed (moving) maximum difference of the signal
- etc.

These linear and nonlinear functions $V_j(\vec{r_i})$ have different variability. The raw signal from the electrode gives rise to high variability due to spiking of the neurons whereas the windowed power and the windowed maximum difference give a smoother behaviour for the CoG trajectories. In the following examples the windowed maximum difference of the signal has been used.

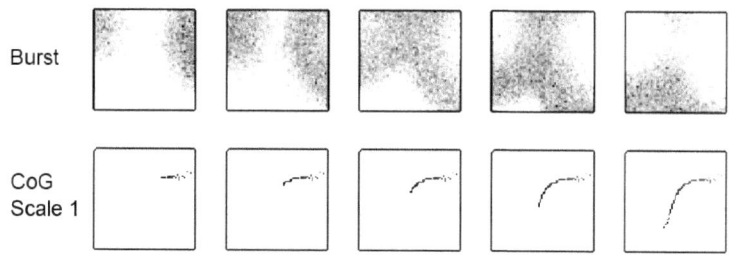

Figure 4.9 : Burst propagation of a neuronal network on a high-density MEA simulated with the INNS of A.Garenne, Bordeaux (top row). Center of Gravity at scale 1 of the same burst (bottom row).

A burst propagation obtained with the INNS is shown in figure 4.9.

As an example, the trajectories of the CoG at four scales for simulated bursts are shown in figure 4.10.

In the following experiment the spontaneous activity of a dissociated hippocampal culture at 27 DIV was recorded (figure 4.11). The wavelet transform was computed for each burst and the CoG was determined at each scale. Each burst was then classified by visual inspection according to the four CoG trajectories. The recordings were performed at DIBE, Genova.

Figure 4.11 also plots the instants of occurrence of each class of burst. The recording length of 120 s is too short to do a reliable statistical burst analysis, however the high-resolution trajectories of the CoG of each burst is one way for a dynamic characterization of the activity of a network. The nature of the identified classes in the recording of figure 4.11 are shown in figure 4.12. The distinct propagation patterns are clearly visible.

In addition to spontaneous activity recordings network properties are also studied by applying specific electrical or chemical stimulation. Several papers have been published in this context showing the learning capacity of a cultured network [9, 10, 112, 185]. The usual technique involves quantifying the network

4.4 Multiscale Analysis - Global vs. Local Activity

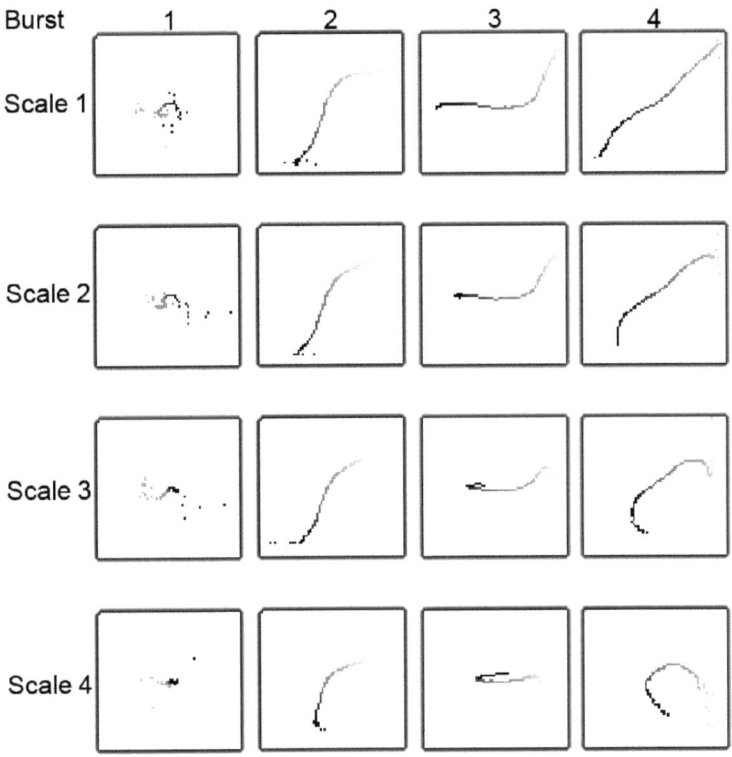

Figure 4.10 : The trajectory of the CoG of the activity from simulated bursts at different spatial resolutions (level 1 - level 4). Four different burst types appeared during the simulation time of 500 ms

activity by means of global statistics, such as PSTH or cross-correlograms, before and after the stimulation. A lasting change in these statistics is then interpreted as a modification of the underlying functional network.

Figure 4.13 and figure 4.14 show the raster plot and the burst classes for the same culture as in figure 4.11 and figure 4.12 after administration of CTZ. Most identified burst classes differ significantly from the ones before CTZ (figure 4.12). These experiments were performed at DIBE.

This experiment does not claim to draw any quantitative conclusion. However, it shows the potential interest of the image-based characterization of network

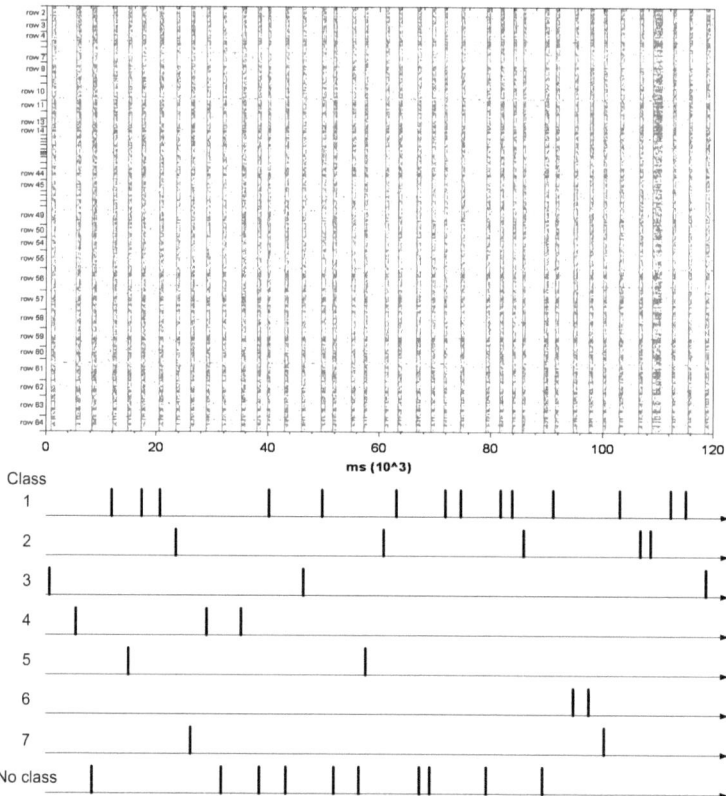

Figure 4.11 : Raster plot of spontaneous activity from a dissociated hippocampal culture at 27 DIV (top). Each burst was analyzed and classified (bottom). Seven different classes could be identified. Bursts that appeared only once during the recording interval were not classified (no class).

dynamics. The multiresolution approach allows focusing on the desired level of detail. In general, as seen in figure 4.12 and figure 4.14 it is necessary to consider all important scales for correct classification of the burst, however in specific cases one or two scales are sufficient for identification.

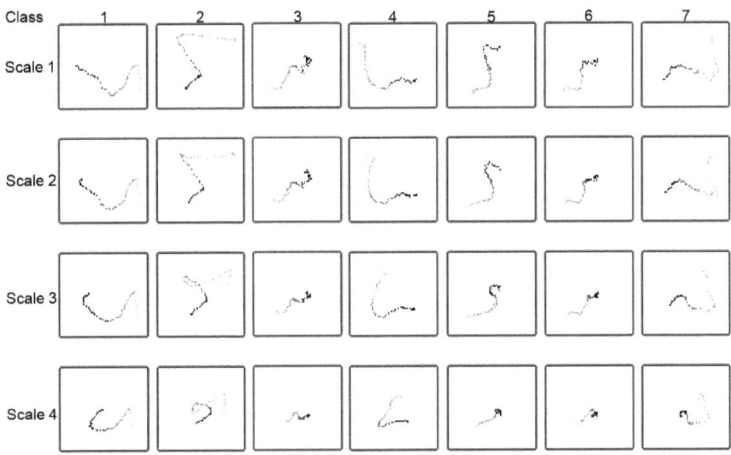

Figure 4.12 : The trajectory of the CoG of the activity from a hippocampal culture (DIV 27) at different spatial resolutions (Level 1 - Level 4). There are seven classes identified within a recording window of 120 s.

4.5 Conclusion

The aim of this chapter was to introduce a few image processing methods that can be investigated in the future for the enhancement and analysis of large-scale high-resolution MEA systems. We stressed the use of the wavelet transform by keeping in mind the idea of real-time/on-line implementations.

The platform developed in this thesis can monitor a neuronal network down to the cellular level. This activity imaging system allows an insight to large networks of neurons at a detail of information that is unprecedented. High-resolution electrophysiological monitoring also enables a new kind of data processing. Algorithms, concepts and ideas from image and video processing fields can be exploited and may lead to complementary sources of methods towards the understanding of physiological phenomena in *in vitro* neuronal cultures.

The nature of signals from high-resolution MEAs is unknown to the best of our knowledge and therefore, an appropriate modelling of a large network coupled

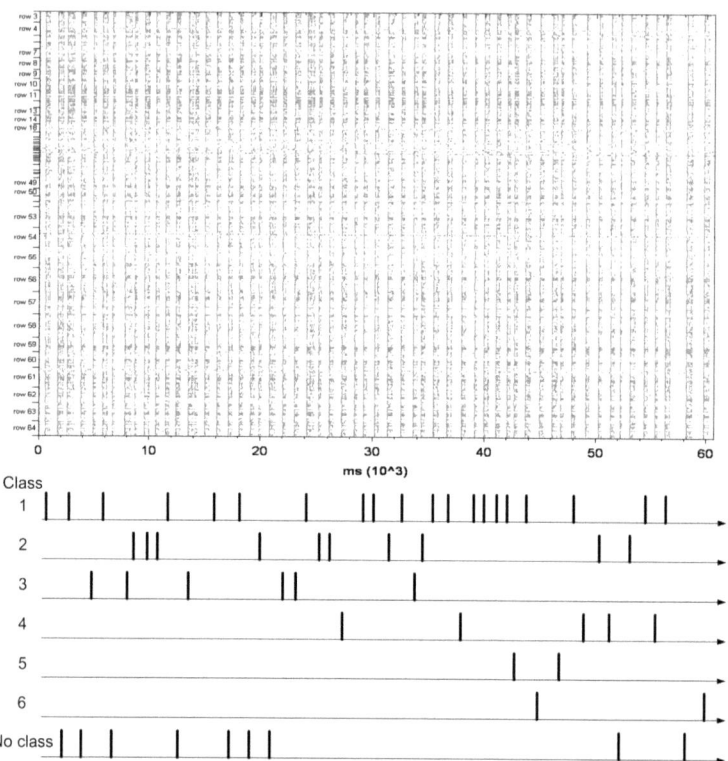

Figure 4.13 : Raster plot of spontaneous activity from a dissociated hippocampal culture at 27 DIV after administration of CTZ (top). Each burst was analyzed and classified (bottom). Six different classes could be identified. Bursts that appeared only once during the recording interval were not classified (no class).

to a large-scale high-resolution platform was necessary for a qualitative and quantitative analysis of new processing methods that take into account the high spatial resolution of the system.

Image and video processing largely benefit from the inherent spatial redundancy of "real-world" images. Redundancy within the network and the codes of neurons were not considered here, however we showed empirically that the measurements from the APS-MEA system provide locally spatially redundant information which can be used subsequently to improve signal detection.

4.5 Conclusion

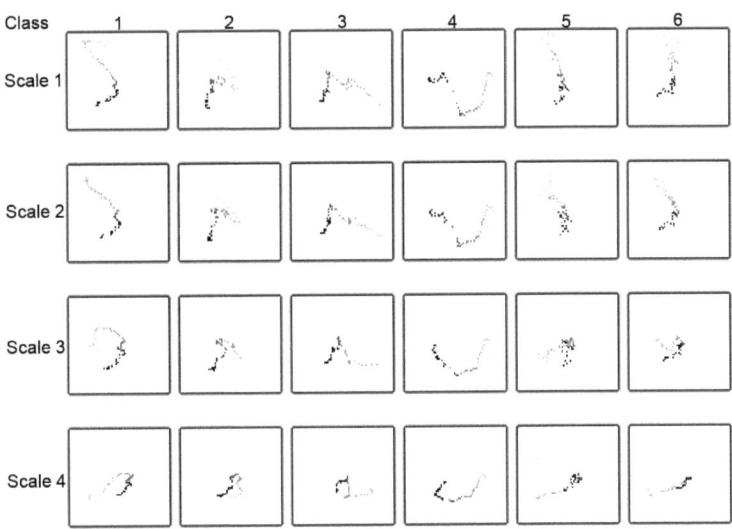

Figure 4.14 : Raster plot of spontaneous activity of the same culture as shown in figure 4.11, after administration of CTZ.

In that sense, two multidimensional denoising methods were tested and compared to the channel-by-channel denoising from section 3.2. One method is based on the spatial decorrelation of multisensor data using the PCA method followed by a channel-by-channel denoising. The second method relies on the spatial decorrelation property of a 2d-wavelet transform, followed by a channel-by-channel denoising. It turned out that the 2d-wavelet transform can efficiently denoise video sequences that consist of well defined spatial objects, however it fails for the enhancement of burst signals due to their lack of significant correlation within a single frame. Independent processing of spatial and temporal information (2d+t) of APS-MEA data does not lead to an improvement of the signal quality unlike this is the case for video sequences of moving objects. Thus, a more complex methodology using mathematically *non-separable* 3d wavelet transforms needs to be considered in the future.

The high resolution feature of the APS-MEA platform opens the door to neuronal network analysis at a level of detail that has not been achieved yet to the best of our knowledge. New types of analysis tools will emerge in order to benefit from the high spatial resolutions of the activity monitoring. In this respect, the multiresolution property of the 2d-wavelet transform is one potential way of characterizing a network at different levels of details. This idea was demonstrated by tracking the propagation of bursting waves across a neuronal dissociated culture. For this purpose, a measure of activity, the center of gravity of the instantaneous activity, was introduced and its evolution at different resolutions was analyzed by a wavelet transform. For instance, complex patterns of activity can be monitored at a coarse level and refined to the desired level of detail when the precision is required.

Contributions

To the best of our knowledge, the following contributions can be considered as original:

- Introducing image/video processing methods and concepts for the enhancement and analysis of signals from the new high resolution electrophysiology platform. In particular, we focused on the

  - Investigation and comparison of spatio-temporal denoising methods, including PCA- and wavelet-based redundancy reduction
  - Multiresolution analysis for characterizing a dissociated neuronal network at different levels of detail
  - Definition of a center of gravity in the wavelet domain for the propagation tracking of network bursts at different levels

Further potential research

- Due to the intrinsic biological correlation of adjacent cells, redundancy shall be explored using *non-separable* 3d-wavelet transforms, which have also recently become increasingly popular in the field of video processing [178] for denoising and motion estimation

Chapter 5

Hardware Implementation for Real-Time Signal Processing

Considering the important computational burden for the analysis of the data from an APS-MEA system, off-line processing on a host computer becomes impractical. In this context, it is a key objective to provide a platform that i) integrates neuronal signal analysis already at the hardware level in order to extract as much as possible significant information at an early stage of the acquisition process and ii) performs all the basic data analysis with a set of simple operations. While more complex mathematical tasks, e.g. the computation of eigenvectors for PCA analysis, are less appropriate for direct hardware implementation, simple operations, such as filtering, thresholding and comparisons are more adequate for efficient hardware-based solutions. As we showed in chapter 3, the approach based on the wavelet transform can handle all fundamental analysis tasks using a set of operations, such as filtering, thresholding, additions and multiplications.

The key operation of wavelet-based processing and the time-critical element is the real-time computation of the transform. Since Mallat came up with a pyramidal filter bank (see figure A.3) to perform a fast DWT [147] many VLSI architectures were proposed for efficient computation of 1d- and 2d-wavelet transforms [186]. Folded architectures [187, 188] use only one low-pass and one high-pass filter and the processing of all levels of the DWT is multiplexed onto these two blocks. Parhi [187] also described a digit-size architecture to improve the hardware utilization efficiency and to reduce routing and interconnection overhead. Vishwanath [189] implements systolic arrays of size N x (J-1) where N is the size of the filter and J corresponds to the maximum level of the DWT and the elements of the array are data processing units, consisting of registers, multipliers and adders. Daubechies and Sweldens [190, 191] introduced the lifting scheme or ladder structure for the efficient computation of the DWT. Recently, several authors [192–195] implemented lifting-based DWTs in VLSI.

Hardware-implemented wavelet transforms for neurophysiological applications were reported by [196] for data compression of 32 channels implantable neuro-prosthetics. The design was targeted to minimize area and power in order to be of practical use in wireless devices. The requirements of hardware-based signal processing for large-scale high-resolution MEAs are different. Consumption and circuit area are not critical. However, the processing speed is fundamental to perform real-time wavelet decompositions of hundreds of electrodes.

Here, we will discuss the design and implementation of a DWT for hundreds and thousands of electrodes. We will present a simple DWT computational core that is configurable up to a filter order of 14. The architecture will be adapted to low-cost FPGAs and is easily scalable to higher number of electrodes. This chapter aims at presenting a key step towards the implementation of a unified hardware-based real-time framework for neuronal signal analysis.

5.1 Architecture

The architecture of the large-scale high-resolution MEA platform is designed in a way to provide in-system hardware resources for real-time computation. The computational units consist of (i) an FPGA and (ii) a RISC image processor. Typically, an FPGA can easily be used for operations such as additions, multiplications and thresholding. Hard-wired functions and inherent parallel processing makes it particularly efficient for those operations. It is not limited to sequential processing as it is the case in software-based implementations (i.e. in a DSP). The current acquisition system includes a Cyclone FPGA from Altera with 20 kLE (Logic Elements)[1], 295 kBit internal RAM and 1 MB of external 100 MHz SRAM.

The fastest way to process a high number of channels is to provide a filter bank for each electrode. This is not feasible for hundreds of electrodes, thus multiplexing of several channels on a few computational units seems more appropriate. In this respect, a folded (i.e. multiplexed) and memory-based architecture was chosen.

An external memory (SRAM of $2^{18} \times 32$ bits) is used to store a number of N-1 past samples of each electrode at each scale that are necessary to compute an Nth order FIR filter. Since the low-pass filter h0 and the high-pass filter h1 will be used by several channels, we will have to store all the c_j of each level for further decomposition whereas all the d_j are continuously sent to the output. A maximum order of 14 for the implementation of Symlet7 filter will be targeted. Thus, 14 samples of each electrode at each level will have to be stored. The samples of all processed electrodes at one instant are organized into a frame (figure 5.1).

---

[1] http://www.altera.com

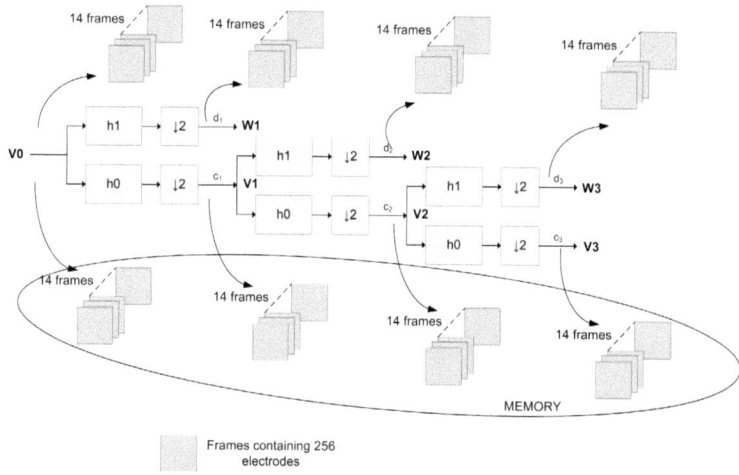

Figure 5.1 : DWT filter bank (3-level) with a memory-based multiplexed architecture. The details W_j are output at each level whereas the approximations V_j are needed for the subsequent decomposition level.

The processing of the entire frame has to be completed before the arrival of a new frame. Since at each level only every other sample needs to be computed (i.e. downsampling), the computation of a sample at each scale is periodic in 2^j, where j is the scale number. This determines the scheduling of the computation of each level using only one common low-pass and high-pass filter unit. The sequencing of each level is shown in figure 5.2.

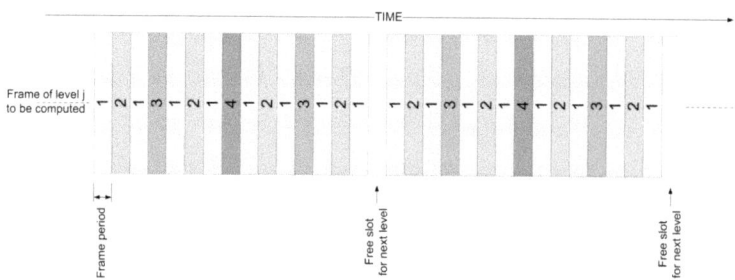

Figure 5.2 : Schedule of wavelet scale j to be computed. The downsampling of the fast DWT guarantees always a free slot. The period for the computation of an update at level j is 2^j.

The bottleneck of this memory-based multiplexed architecture is the speed of

5.1 Architecture

the READ and WRITE access operations from and to the SRAM. Since the full word length of 32 bits will be used for the internal representation of intermediate fixed-point results, only one coefficient c_j of one electrode can be read during one clock cycle. Therefore, 14 clock cycles are needed to compute the next levels c_{j+1} and d_{j+1} and all the electrodes contained in one frame need to be processed sequentially. Moreover, the memory has to be accessed in blocks for READs and WRITEs in order to reduce the mode setup time between a READ and a consecutive WRITE. Thus, the maximum number of electrodes that can be processed with the current memory bandwidth limitation is 256. In fact, with a minimal frame rate of 15 kHz for 256 electrodes/frame and a clock frequency of 75 MHz the FGPA can allocate 5000 cycles between two signal frames for the computation of a new frame of wavelet coefficients. 3580 ($= 14 \cdot 256$) cycles are required for the processing of one frame leaving the remaining clock cycles to store the incoming samples and the approximation coefficients c_{j+1} of the next level. 12 bits of the 18-bit address are used to access the 14 samples of the 256 electrodes. The remaining 6 bits provide room for 64 wavelet decomposition levels (figure 5.3).

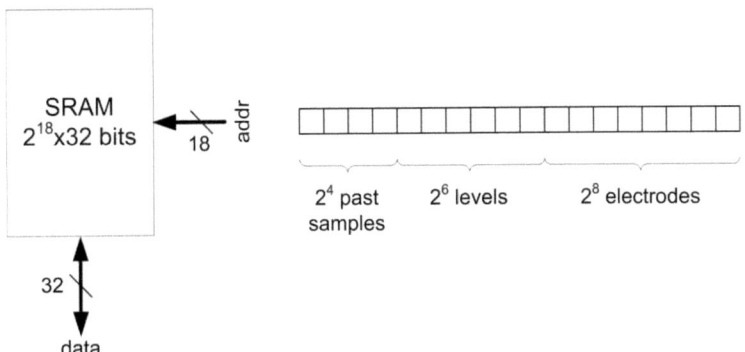

Figure 5.3 : The 18 bit address is used to access all the maximally required approximation coefficients at each electrode

As mentioned above, the asynchronous SRAM can perform consecutive READ accesses each cycle for clock frequencies up to 100 MHz. However, it takes two cycles for a correct WRITE operation at this speed. Therefore, READ and WRITE operations should not be performed in interleaved cycles, but block wise instead. For this reason the processing time between the arrivals of two frames is divided into three distinct phases:

- Write the arriving frame in the SRAM

- Read the SRAM for the computation of the wavelet transform of all channels

- Write the coefficients c_{j+1} of all electrodes into the SRAM

Figure 5.4 shows the architecture and data flow of the wavelet processor. The central component is the sequencer that manages the timing of the READ and WRITE accesses to the different memory blocks that store the intermediate results and the past samples of each electrode for each scale. The address generator encodes the addresses according to the scheme in figure 5.3.

The filter configuration unit on the right of figure 5.4 receives the variable filter coefficients from the host computer. The internal precision of these coefficients is set to 24 bits. The computational units (CU) h0 and h1 perform the filter operations of each electrode channel. It contains *unsigned* 24 bits x 18 bits multipliers, decomposed into a 3-level-pipeline. For the ease of implementation, an *unsigned* multiplication is used. However, wavelet filters can have *signed* coefficients and therefore, *signed* multiplication has to be carried out by using *unsigned* operations only.

Two *signed* numbers x and h of n_x and n_h bits, respectively, can be expressed by two *unsigned* numbers x' and h' as follows:

5.1 Architecture

$$x = x' - c_1 \qquad \text{signal samples} \qquad (5.1)$$

$$h = h' - c_2 \qquad \text{filter coefficients} \qquad (5.2)$$

A filtering operation includes a sum of multiplications:

$$\sum_{i=1}^{n} x_i \cdot h_i = \sum_{i=1}^{n} (x'_i - c_1) \cdot (h'_i - c_2)$$

$$= \sum_{i=1}^{n} (x'_i \cdot h'_i - x'_i \cdot c_2 - c_1 \cdot h'_i + c_1 \cdot c_2)$$

$$= \sum_{i=1}^{n} x'_i \cdot h'_i - c_2 \cdot \sum_{i=1}^{n} x'_i - \sum_{i=1}^{n} (c_1 \cdot h'_i - c_1 \cdot c_2) \qquad (5.3)$$

One observes that a sum of the products of two *signed* numbers corresponds to the sum of the products of two *unsigned* numbers corrected by two terms. The second of these correction terms is a constant. It only depends on the filter coefficients h'_i and the positive offsets c_1 and c_2 and it has to be computed once for a given filter. The first correction term also depends on the samples x'_i and needs to be computed for every multiplication. However, since in the present case h is a 24 bit number, thus $c_2 = 2^{23}$, the first correction term can be computed by a sum of left-shifted x'_i (i.e. multiplication by 2^{23}).

Currently, only two filter units (i.e. h0 and h1) are needed for the processing of the 256 electrodes. The logic blocks for the filters implemented in an FPGA with 20 kLE occupy about 10 % of the total resources. Additional filter stages for the processing of more electrodes can easily be implemented. Thus, one can conclude from the above discussion that more electrodes can be processed by increasing the bandwidth of the available memory. By adding n SRAMs in parallel, n electrodes can be transformed in parallel. Therefore, the total number of electrodes can be increased by a factor n with $n \leq 8$ in the case of the current low-cost Cyclone 20 kLE from Altera.

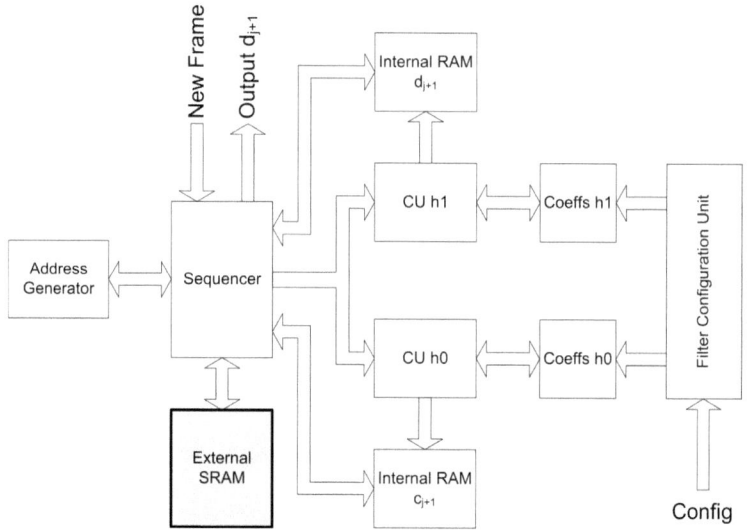

Figure 5.4 : Data flow of wavelet processor. The sequencer manages the timing of each operation. The computational units (CU h0 and CU h1) perform the arithmetic operations for the filter h0 and h1. The filter configuration unit allows flexible programming of the filter coefficients h0 and h1. The internal RAMs store temporarily the results for each wavelet level, before accessing the external SRAM for long-term storage.

5.2 Validation

The development and simulation was performed with HDL Designer and ModelSim from Mentor. The final validation takes place by feeding an input signal to the wavelet processor.

Figure 5.5 shows the comparison of two DWTs of a known input signal. The DWT decomposition on the left is obtained from the hardware-implemented filter bank. On the right, one can see the DWT of the same signal obtained using the Wavelet Toolbox of Matlab. One notices that there is a good match between the software-based and the real-time hardware-based solution.

For the purpose of demonstration, figure 5.6 shows the real-time DWT of a burst from a rat's dissociated cortical culture. This real-time wavelet transform can

5.2 Validation

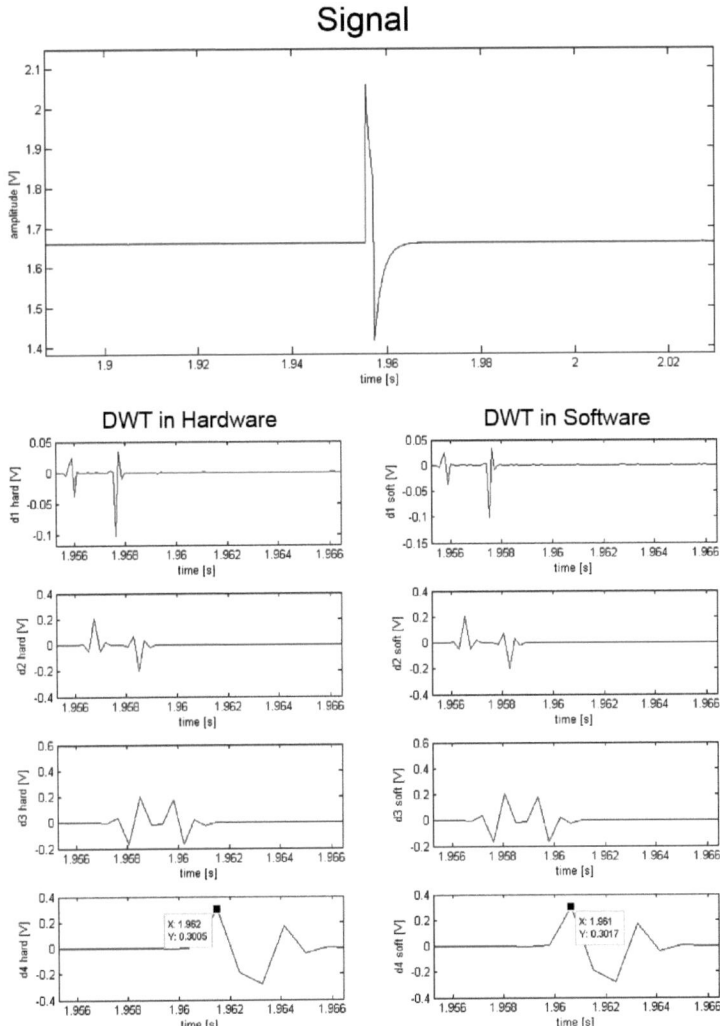

Figure 5.5 : Validation of wavelet processor. A known signal is fed to the wavelet processor and the hardware-based (real-time) DWT (left) is compared with the DWT obtained from a Matlab-function (right).

easily be extended to real-time spike detection and sorting by using the algorithms presented in chapter 3, either implemented on the FPGA in hardware, or occasionally implemented on the RISC processor of the acquisition system in software. For the sake of simplicity, the FPGA can be configured to implement thresholding in the wavelet domain for a quick measure of activity.

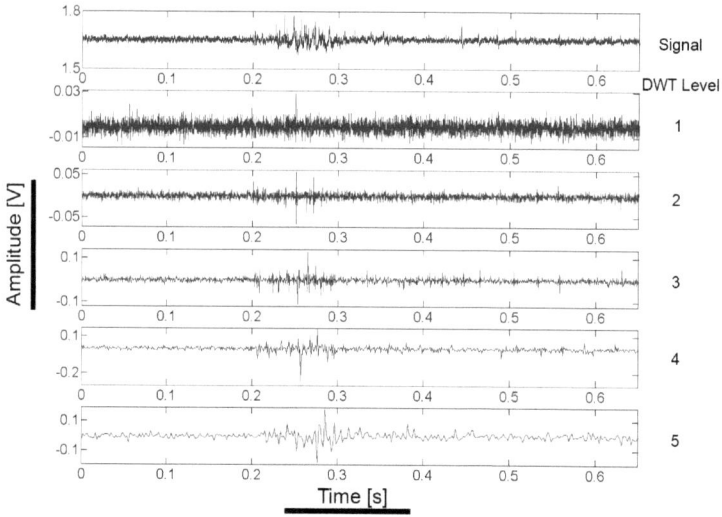

Figure 5.6 : Real-time wavelet decomposition of a burst. Only one electrode is shown, together with the first 5 scales. Higher scale gives details at a lower resolution.

5.3 Application: Activity Monitoring on High-Density MEAs

Real-time activity monitoring of a large number of electrodes is an important tool for the experimenter. With conventional systems up to hundreds of recording sites host computers are usually sufficient for detection and displaying electrophysiological activity. Manual navigation across the recording sites is man-

5.3 Application: Activity Monitoring on High-Density MEAs

ageable. The problem arises when several hundreds or thousands of electrodes shall be monitored for activity at the same time. One can refer to chapter 3 for spike detection and sorting when the precise timing of individual neurons is required. However, for a general overview of the activity, single neuron contributions are not necessarily important and one can rather define a density of activity, which expresses the local strength of activity. Furthermore, a continuous display of on-line or real-time spike detection can not be visually perceived by the experimenter due to the high frame rate of the images (e.g. several kHz). Running spike rates are more appropriate to represent network dynamics on the fly.

We introduced the WTMM in section 3.3 for the detection of spikes. Mallat showed in [140] that the discontinuities of a signal can be determined by locating the points of convergence of the WTMMs from higher to lower scales j. The more WTMMs are found inside a binning window, the more singularities, i.e. spike events, are present within the same window. Thus, one can count the number of WTMMs at the output of the wavelet processor in order to get an activity representation over a large amount of electrodes. The idea is illustrated in figure 5.7.

The method was tested with a simulated reference signal using the model of section 4.1. Figure 5.8 compares the activity representation obtained from a conventional running spike rate (top) and the one obtained from the binning of WTMMs (bottom). It can be noticed that the activity monitoring with WTMMs tends to be smoother. WTMMs also occur at higher scales and therefore they have an averaging effect on the WTMMs from lower scales.

Figure 5.9 shows a sequence of activity frames of a burst in a dissociated cortical culture from rat. It has to be emphasized that the WTMM activity monitor-

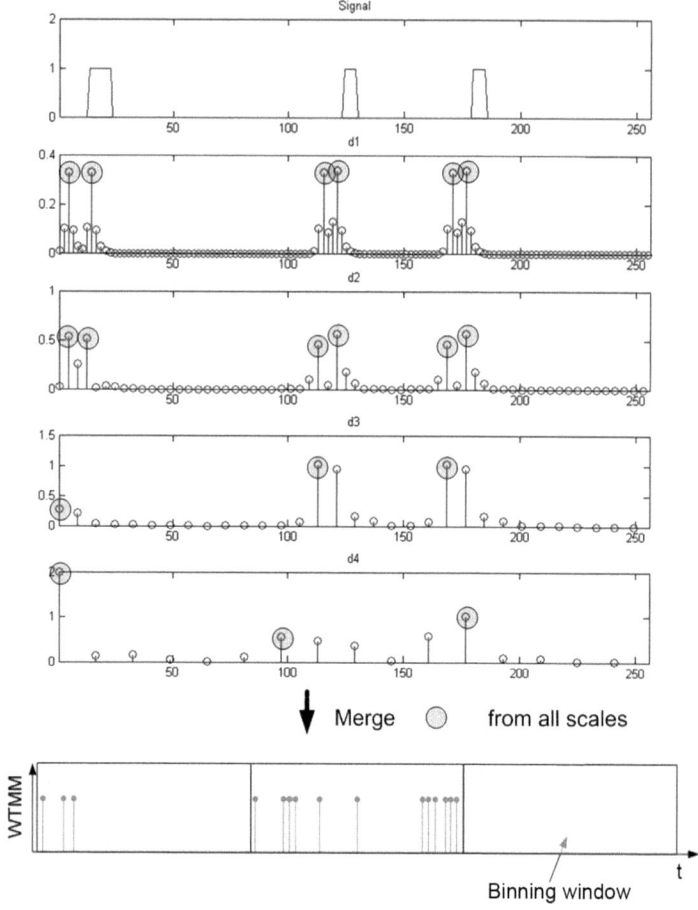

Figure 5.7 : The wavelet transform modulus maxima (WTMM) are determined at each level d_j of the DWT. Counting the WTMMs from all scales within a window gives a local measure for the number of singularities in this window.

5.4 Conclusion

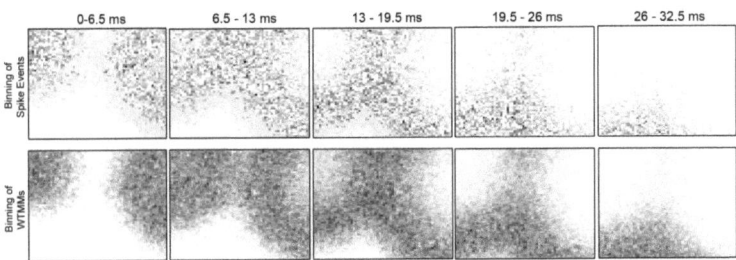

Figure 5.8 : Comparison of activity detection in a simulated burst by binning detected spike events (top) and by binning WTMMs (bottom). Bin size 50 samples (i.e. 6.5 ms).

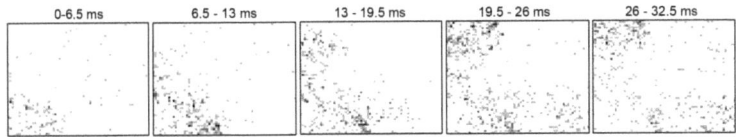

Figure 5.9 : Activity monitoring (bin size 6.5 ms or 50 samples) using WTMM during a burst in a dissociated cortical culture from rat.

ing has its purpose in the on-line and real-time visualization of an on-going experiment.

5.4 Conclusion

The strategy for real-time implementation of the analysis tasks discussed in chapter 3 is the migration to hardware. At that level, the computation of a DWT can be performed efficiently by a set of filters. Successively, simple operations on discrete wavelet coefficients enable spike detection and clustering for spike sorting, but also support convenient tools for data visualization.

The low complexity of all the operations of the wavelet-based algorithms discussed in chapter 3 enables the full integration in off-the-shelf hardware. A real-time wavelet processor on 256 channels was implemented on the current high-density MEA platform to delineate the methodology towards a full real-time signal processing framework for high-density MEA acquisition. We discussed its architecture on an FPGA and showed that it can easily be extended to larger number of electrodes by scaling memory resources. We also showed that it is feasible to handle the data from high-density MEAs in real-time opening new perspectives for the on-line data visualization, denoising, compression and analysis.

Contributions

To the best of our knowledge, the following contributions can be considered as original:

- Hardware-based strategy for real-time analysis of neurophysiological signal
- Wavelet processor architecture and implementation for high-density microelectrode arrays
- Real-time activity visualization based on wavelet transform modulus maxima

Further potential research

- Sweldens [190] suggested a lifting scheme for more efficient computation of the FFT. Although speed of the CUs is not the critical issue in this work, alternative architectures should be anticipated in order to save logic resources
- Extend architecture to work with wavelet packets in order to investigate real-time adaptive schemes for wavelet decomposition and benefit from the expected increase in detection and denoising performance. Alternatively, the configurable architecture also allows the implementation of matched

filters for the detection of neuronal signals

Chapter 6

Conclusion and Perspectives

The overall aim of this work was to realize a MEA-based platform with largely enhanced functional chatactersitics with respect to the spatio-temporal resolution, on-line data visualization and real-time signal analysis. The main objectives were to (i) develop a high-resolution MEA platform, (ii) provide a methodology for real-time signal and data processing, (iii) introduce image-based processing concepts and (iv) validate the platform with various cell cultures.

All achievements and contributions of this work have been discussed in chapters 2, 3, 4 and 5 and they are separately listed in the conclusion section of each chapter. Here, they are summarized as follows:

- A large-scale and high-resolution MEA platform was designed and implemented. The system contains a CMOS-based array of 4096 electrodes on an area of $7\,mm^2$. The electrode size is $21\,\mu m \times 21\mu m$ with an electrode-to-electrode distance of $21\,\mu m$. The architecture and implementation was inspired by concepts from the fields of image / video acquisition and processing. The system currently generates about $500\,Mbit/s$ of electrophysiological data.

- It is intrinsically necessary to provide a methodology for real-time signal processing, which is required for practical use of systems with large numbers of channels, and that is adapted to a high amount of data. The proposed strategy includes migration of software-based neurophysiological analysis tasks to hardware-based data processing. In general, hardware-implemented analysis can operate more efficiently than software-based solutions, as it is strongly dependent on the application and therefore less flexible than software-based approaches.

- In order to maximize the number of signal processing tasks to be implemented in hardware, a general-purpose tool was required. Hence, a wavelet-based framework was investigated. This framework relies on the computation of a fast wavelet transform that can be efficiently implemented using filter banks. Subsequently, we showed that all necessary tasks for neurophysiological signal processing can be performed within this framework. This includes (i) denoising, (ii) spike detection and (iii) spike sorting.

- A new spike sorting algorithm was proposed. Based on the wavelet coefficients that can be computed by a fast wavelet filter bank, different spike types can be discriminated in the wavelet domain. This spike sorting algorithm also plays a fundamental role to remove interferences and induced noise from digital electronics. Due to the use of fast filter-based wavelet transforms with a complexity of $O[N]$ the concept can be transferred to a real-time implementation on hardware.

- The high spatial resolution of the platform enables the use of image / video-based processing methods in the field of electrophysiology. Following concepts were introduced and discussed:
    - The wavelet-based framework was extended to a new 2d multiscale approach for the high-level analysis of neuronal networks. The multires-

olution property of the wavelet transform enables network analysis at different levels of abstraction, ranging from details at the cellular level to more behavioural characterization at the network level. This idea was demonstrated by introducing a simple parameter for the analysis and characterization of bursts across a neuronal network. For that purpose, the center of gravity (CoG) was introduced in order to track the propagation of a burst at different levels of resolution.

- It was shown that signals from different electrodes correlate in space. This redundancy can be used to enhance signals using spatial decorrelation. PCA-based signal decorrelation was shown to improve the SNR of simulated signals from APS-MEAs. As opposed to the PCA-method the intrinsic spatial decorrelation property of a 2d-wavelet transform does not improve the signal quality if temporal correlation between frames is not included in the statistical processing.

- A real-time wavelet processor was implemented and validated on a low-cost FPGA for the processing of 256 electrodes. The proposed *folded* architecture is easily scalable to larger numbers of electrodes by increasing the data bandwidth between processor and RAM. This implementation delineates the step towards an efficient real-time analysis of high-density MEAs consisting of thousands of electrodes.

- The platform was validated with a variety of dissociated cell cultures including cardiomyocytes as well as hippocampal and cortical neurons. Chemical stimulators were used to induce changes in the activity patterns in order to demonstrate the operation of the system.

One final aim is to use the system under a wide range of experimental conditions for investigating relations between local phenomena in electrogenic cells, such as propagation effects or synaptic changes and network characteristics, such as overall dynamics (e.g. firing rate, network bursting), information processing mechanisms (i.e. population code) and network plasticity. Wavelet-based neurophysiological signal processing on high-resolution MEAs also opens various interesting fields for further research. On the one hand, long-term experiments (for several days) with extraction of biologically significant network parameters at different spatial and temporal scales could be performed without (i) excessive storage requirements for the large amount of raw data and without (ii) long-lasting subsequent off-line processing. On the other hand, given the high spatial resolution of the APS-MEAs, methods for the extraction of interesting activity patterns at both the cellular and network levels inspired from the image processing field can be developed.

The results achieved show the feasibility of large-scale, high-density MEA neurointerfaces that can *concurrently and continuously* record from thousands of electrodes. The platform was validated with a set of cells ranging from cardiomyocytes and hippocampal neuronal signals to low-amplitude spikes from cortical neurons. All cell types were extracted from embryonic rats. Real-time processing of the large amount of data was proposed by outsourcing analysis tasks to hardware. A fast wavelet transform framework enables hardware-based algorithms for the processing of thousands of neurophysiological signals. In this respect, the platform was designed to contain sufficient hardware resources (i.e. FPGA) for the integration of specific computing tasks.

The large number of densely integrated electrodes also leads to an alternative two-dimensional, i.e. image-like, representation of information. Hence, analysis and characterization of neurophysiological data can benefit from concepts of the image/video field and were introduced in this thesis.

The new representation of neurophysiological data in terms of sequences of images strongly emphasizes on potential multidimensional analysis and investigation of biological information. The high spatio-temporal resolution of the data potentially enables new insights into the world of complex neuronal networks. Higher-dimensional search for biologically relevant patterns was not addressed in this thesis, however multidimensional multiresolution analysis, as proposed in this work, could lead to a new approach to uncover some of the underlying biological principles of neuronal networks. In that sense, pattern extraction and "object" analysis based on *non-separable* (and/or complex) 3d-wavelet transforms could provide a good starting point for future work in neurocomputational research.

The wavelet-based framework presented here, and including both single-channel and image processing, also provides a first step towards the development of an efficient computational platform. Hardware-based analysis using one-dimensional wavelets can be scaled to higher dimensions to address new families of (multivariate and/or multidimensional) techniques not yet applied to neuronal signals.

Appendix A

Mathematical Background - Wavelets

A.1 Introduction

Wavelet theory spans a wide class of signal decompositions. The idea of this chapter is to give a short overview of fundamental concepts that are used in wavelet signal processing systems. Some of those concepts will be used in the context of this thesis and it is therefore illustrative to review the basic relations between the different wavelet systems. It will become clear why the wavelet transform is an appropriate tool for multiscale analysis and why its concepts enables a number of interesting research topics in the field of processing and analysis of high-density MEA systems.

A.1.1 Wavelet Transform

The Continuous Wavelet Transform (CWT), as defined in [143], is expressed as:

$$WT(u,s) = \frac{1}{\sqrt{s}} \int_{-\infty}^{+\infty} f(t) \cdot \Psi^*\left(\frac{t-u}{s}\right) \cdot dt = f(u) * \overline{\Psi}_s(u) \qquad (A.1)$$

$$\overline{\Psi}_s(u) = \frac{1}{s}\Psi^*\left(-\frac{t}{s}\right) \qquad (A.2)$$

and can be interpreted as the degree of similarity of a signal f(t) and the analyzing wavelet Ψ(t) at the translation u and scale s. The nature of a non exhausive set of widely used wavelets Ψ(t) can be seen in figure A.1. Intuitively, one understands that the analyzing wavelet Ψ(t) should be as close to the signal f(t) as possible to get the maximum value for WT(u,s). For that reason Daubechies4, Symlet5 and Symlet7 of figure A.1 seem to be good choices for the analysis of extracellular signals (compare with figure 1.2).

Unlike the Fourier transform, the basis functions of a wavelet transform (WT) can have finite support and are therefore well suited to represent event-like signals. Thus, the WT is particularly adapted for the local analysis and characterization of electrophysiological signals. In fact, it can be shown that the wavelet transform can compact a wide class of real-world signal to a few coefficients in the transformed domain by choosing an appropriate basis system [197]. Therefore, denoising and compression techniques perform particularly well in the transformed domain since the signal is mapped into a few high amplitude coefficients whereas random noise is spread over a wide range of small coefficients. A thresholding in the transformed domain can thus separate noise from a signal even though their Fourier spectra may overlap and where a Fourier-based signal processing approach using linear filters would fail. Since the scale s can be arbitrarily chosen in equation A.1, the wavelet transform can also be seen as a multiresolution analysis of a signal f(t).

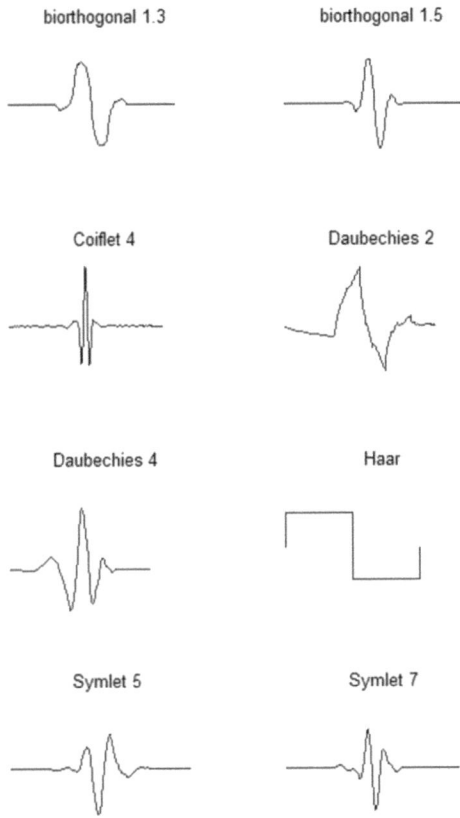

Figure A.1: Waveforms of a few widely used wavelets

A.2 Discrete Wavelet Transform

The Fourier transform $\Psi_s(\omega)$ of $\overline{\Psi}_s$ is

$$\Psi_s(\omega) = \sqrt{s} \cdot \Psi^*(s\omega) \quad (A.3)$$

Equation A.3 states that the (continuous) Wavelet Transform at different scales s can in general be obtained from dilated versions of a band-pass filter $\Psi^*(w)$.

The CWT has to be discretized in order to be practical within numerical algorithms. One could take values at discrete t, u and s for the functions f(t) and $\Psi^*(t)$ and approximately compute equation A.1 using numerical integration. It turns out that this approach is not very efficient in terms of the number of required operations [146]. If we want to get the wavelet coefficients at increasing scales s and keep in mind that the WT at that scale is the result of a band-pass filtering step with decreasing maximum frequency dilation then we would intuitively conclude that the continuous WT(u,s) can be discretized in u on a sampling grid that depends on the scale s. Thus, equation A.1 could be sampled on an octave base as follows and the discretization of the integral (with $\Delta t=1$) leads to w, the discretized wavelet series:

$$WT(s,u) \hat{=} w(2^j, 2^j \cdot k) \hat{=} \frac{1}{\sqrt{2^j}} \sum_n f(n) \cdot \Psi^*\left(\frac{n}{2^j} - k\right) \qquad (A.4)$$

A.2.1 Orthogonal Wavelet Transform

An orthonormal multiresolution analysis of the signal f(t) lying in the space V_{j-1} can be expressed with the two orthogonal subspaces V_j and W_j (figure A.2) as:

$$f(t) = \sum_k c_j(k) \cdot 2^{-j/2} \rho\left(\frac{t}{2^j} - k\right) + \sum_k d_j(k) \cdot 2^{-j/2} \varphi\left(\frac{t}{2^j} - k\right) \qquad (A.5)$$

with $f(t) \in V_{j-1}$ where $\underset{k}{Span}\{2^{-j/2} \cdot \rho(\frac{t}{2^j} - k)\}$ defines the space V_j and $\underset{k}{Span}\{2^{-j/2} \cdot \varphi(\frac{t}{2^j} - k)\}$ defines the complementary space W_j with

$$V_{j-1} = V_j \oplus W_j \qquad (A.6)$$

A.2 Discrete Wavelet Transform

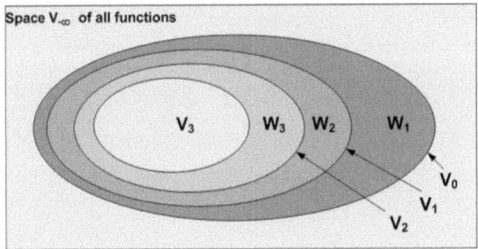

Figure A.2 : The detail space W_j and the approximation space V_j are orthogonal. Both spaces together span the space V_{j-1} of the next higher level.

ρ is the scaling function and φ corresponds to the wavelet function (mother wavelet) of the discrete wavelet system. The coefficients c_j and d_j from equation A.5 are called the discrete wavelet transform (DWT) of the function f(k). Since $\rho_{j,k} = \rho(\frac{t}{2^j} - k)$ and $\varphi_{j,k} = \varphi(\frac{t}{2^j} - k)$ are orthogonal bases of the subspace V_j and W_j respectively [144], the decomposition of equation A.5 is a non-redundant representation of the function f(k).

Equation A.5 can also be interpreted as a multiresolution decomposition of the function f(t) into signal components at different scales j. An increasing scale j gives the details at a lower resolution 2^{-j}. Equation A.5 expresses the analysis steps where a signal f(t) is continuously split into its lower resolution components. At each step the signal consists of approximation coefficients $c_j(k)$ and detail coefficients $d_j(k)$ that are needed to build the signal at the next higher resolution $2^{-(j-1)}$. Consequently, the signal is decomposed from a high resolution to a low resolution at scale j_n:

$$f(t) = \sum_n c_{j_n}(k) \cdot 2^{-j_n/2} \rho(\frac{t}{2^{j_n}} - k) + \sum_k \sum_{j=j_n}^{-\infty} d_j(k) \cdot 2^{-j/2} \varphi(\frac{t}{2^j} - k) \quad (A.7)$$

Equation A.6 states that a function lying in the space V_j also lies in the space V_{j-1}. Therefore, the scaling functions $\{\frac{1}{\sqrt{2}}\rho(\frac{t}{2} - k)\}_{k \in \mathbb{Z}}$ spanning the space V_1

can be expressed in the space V_0, spanned by $\{\rho(t-k)\}_{k\in\mathbb{Z}}$:

$$\frac{1}{\sqrt{2}}\rho(\frac{t}{2}) = \sum_k h(k) \cdot \rho(t-k) \qquad (A.8)$$

Since $W_1 \subset V_0$ the corresponding relationship for the wavelet function $\varphi(t)$ can be found as:

$$\frac{1}{\sqrt{2}}\varphi(\frac{t}{2}) = \sum_k g(k) \cdot \varphi(t-k) \qquad (A.9)$$

Equation A.8 and A.9 are the fundamental recursive equations for the multiresolution analysis of a signal $f(t) \in L^2(\mathbb{R})$, where $L^2(\mathbb{R})$ is the class of all functions that have a finite, well-defined integral of the square [144].

Equation A.5 and A.7 indicate that by keeping the details $d_j(k)$ at each step no information is lost and therefore the signal can be reconstructed from the set of its lowest resolution approximation coefficient $c_{jn}(k)$ and all the detail coefficients $d_{j(k)}$ down to j_0 where $j_0 < j_n$.

Orthogonal wavelet transforms are very efficient in terms of the number of computational operations for a given set of samples to be processed. A function $f(t) \in V_{j-1}$ can be represented by the approximation coefficients c_j and the detail coefficients d_j. By using equations A.7, A.8 and A.9 we can get a set of recursive equations that express the coefficients c_j and d_j at the next higher scale j as a function of the approximation coefficients c_{j-1} from the lower scale (i.e. finer resolution) j-1:

A.2 Discrete Wavelet Transform

$$c_j(k) = \sum_m h(m - 2k) \cdot c_{j-1}(m) \qquad (A.10)$$

$$d_j(k) = \sum_m g(m - 2k) \cdot c_{j-1}(m) \qquad (A.11)$$

This multiresolution algorithm was first expressed by Mallat [147] and is commonly referred to as orthogonal discrete wavelet transform (DWT). It needs to be initialized and if the signal is sampled at a sufficiently high rate then the samples $f(k)$ of the original signal $f(t)$ can be set equal to the fine scale approximation c_0. Interesting enough that equations A.10 and A.11 decomposes a signal f(t) into different subbands and that this set of equations can be interpreted as an analysis filter bank shown in figure A.3. After each low-pass filter g(k) and high-pass filter h(k) the signal is subsampled by 2. This leads to a constant number of output samples and therefore the overall complexity of the DWT is O[N]. Due to the subsampling at each stage the number of operations is independent from the number of scales to be computed.

Equations A.10 and A.11 also relate wavelet systems to the well known theory of filter banks [198]. It is worth mentioning that the basis functions, both the scaling function $\rho(t)$ and the mother wavelet $\varphi(t)$, are not directly used to compute the DWT, however they govern the characteristics of the filters h(k) and g(k) through the multiresolution equations A.8 and A.9.

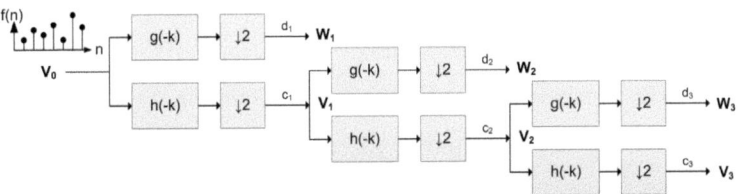

Figure A.3 : Filter bank for the fast computation of the orthogonal DWT

Orthogonal wavelet systems are known to be a robust and non-redundant representation of a signal. Parseval's energy partition theorem is valid and it produces very few coefficients for many classes of signals (i.e. energy compaction property). However, one major drawback is its shift-variance. By looking at equation A.4 one can see that a discrete function f(n) with its discretized wavelet series $w(2^j, 2^j \cdot n)$ does generally not produce the same time-shifted wavelet series $w(2^j, 2^j \cdot (n-r))$ for a shifted function f(n-r). This is because the translation variable is sampled as a function of the scale, i.e. at an increased sampling period with an increased scale. This property is very critical in pattern recognition tasks and classification problems (i.e. spike classification) since the feature set that serves to identify the signal type has to be independent of any shift of the signal in the time domain.

A.2.2 Redundant Wavelet Transform

A frame is a set of vectors that represent a signal f(t) from its inner products with the frame vectors [147]. The set of vectors does not have to be independent, i.e. a basis, but it has to be a spanning set. Such a frame can be constructed, for instance, by taking the orthogonal basis vectors from equation A.5 and augment this orthogonal spanning set by the shifted orthogonal basis vectors. The resulting representation can also be interpreted as an average of all the shifted orthogonal wavelet transforms representations and gives a highly redundant and translation-invariant representation of the signal [199].

Another approach to the understanding of shift-(in-)variance follows from the definition of the continuous wavelet transform (equation A.1). Instead of sampling the WT along the scale and along the translations at multiples of 2^j for scale j, as in the case of the orthogonal wavelet transform, one can sample the

A.2 Discrete Wavelet Transform

WT only along the scale, but not along the translation u. This leads to the definition of the *dyadic* wavelet transform:

$$WT(2^j, u) = \int_{-\infty}^{+\infty} f(t) \frac{1}{\sqrt{2^j}} \Psi^*(\frac{t-u}{2^j}) \cdot dt = f * \bar{\Psi}_{2^j}(u) \qquad (A.12)$$

Similar to the general CWT from equation A.1, the dyadic WT can be discretized. This leads to an undecimated discrete wavelet transform that can be efficiently computed by an algorithm also known as "algorithme à trous" [200]. Its shift-invariance is traded against a redundant, overcomplete representation. The "algorithme à trous", shown in figure A.4 , has a computational complexity of $O[N \cdot log(N)]$. Intuitively, one can also argue that shift-invariance is restored by omitting the downsampling at each level of the filter bank in figure A.3. However, it is also clear that we get an overcomplete representation (i.e. oversampled) because we keep more samples at the output of the filter bank than it is basically required to perfectly reconstruct the signal. In the remainder of this thesis we will refer to shift-invariant discrete wavelet transforms as SIDWT.

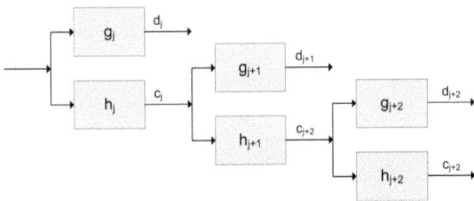

Figure A.4 : Fast dyadic wavelet transform, known as "filtre à trous". The filters g_{j+1} and h_{j+1} of the next stage are obtained by inserting 2^{j-1} zeros (i.e. holes ="trous") between each sample of the original filter g and h.

A.2.3 Wavelet Packets

The "classical" DWT is computed with a fast filter bank implementation as shown in section A.2.1. Only the approximation space is further decomposed for the next level iteration. However, Coifman [201] showed that a more flexible frequency tiling can be obtained by also allowing a recursive splitting of the detail subbands. It was shown that the detail subband W_j can also be decomposed into orthogonal subspaces $W_{0,j+1}$ and $W_{1,j+1}$. A complete wavelet tree up to level 3 is shown in figure A.4. Any tree is called an admissible tree if each node has either 0 or 2 children (figure A.4, left, dark gray shaded blocks). All the subspaces spanned by each branch of an admissible tree recover the original signal space V_0. An admissible tree structure is called a wavelet packet system [144]. One of its advantages resides in the fact that the structure of the signal bases can be adapted to the frequency characteristics of the input signal since the tiling of the frequency plane can be optimized to best match the signal. The upper bound of the computational complexity of the entire wavelet packet tree is $O[N \cdot \ln N]$, in contrast to $O[N]$ for the classical orthogonal DWT of section A.2.1, and therefore, similar to the complexity of the FFT.

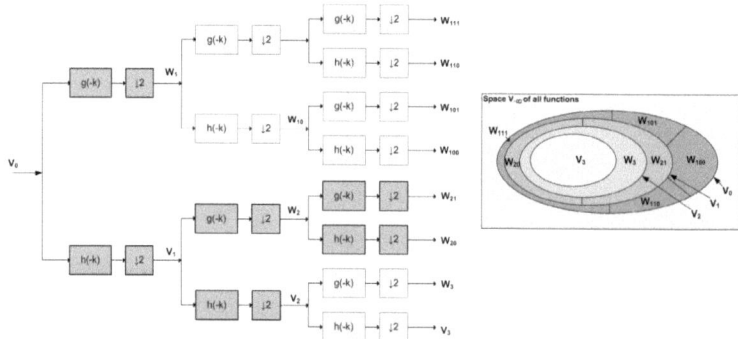

Figure A.5 : The entire wavelet packet tree for three levels of decomposition (left). The tree allows the choice of a basis system other than the conventional wavelet transform. This enables a very flexible frequency tiling (right). One example of a valid and non-redundant decomposition is shown in dark gray.

A.3 2d-Wavelet Transform

A straightforward extension of one-dimensional wavelet systems leads to two-dimensional wavelet transforms by taking the scaling function $\rho(t)$ and the mother wavelet $\varphi(t)$ analogous to equation A.5. From these two functions, four *separable* functions of x and y are created as follows:

$$\rho(x,y) = \rho(x) \cdot \rho(y)$$
$$\varphi^H(x,y) = \varphi(x) \cdot \rho(y)$$
$$\varphi^V(x,y) = \rho(x) \cdot \varphi(y)$$
$$\varphi^D(x,y) = \varphi(x) \cdot \varphi(y) \tag{A.13}$$

These functions measure variations (i.e. edges) along different directions of an image: $\varphi^H(x,y)$ in the horizontal, $\varphi^V(x,y)$ in the vertical and $\varphi^D(x,y)$ in the diagonal direction. The four functions are *separable* since they are the product of two one-dimensional functions. They are a subset of the more general class of two-dimensional wavelet functions that also contain *non-separable* wavelets. *Separable* two-dimensional wavelet systems, such as given in equation A.13, can easily be computed by taking a 1d-wavelet transform on the rows followed by a 1d-transform on the columns of the resulting image. One filter stage of a 2d-*separable* wavelet system is shown in figure A.6.

All the properties for 1d-wavelet transforms discussed above, i.e. orthogonal DWTs, redundant wavelet transforms and wavelet packets, etc., can be similarly applied to 2d-wavelets.

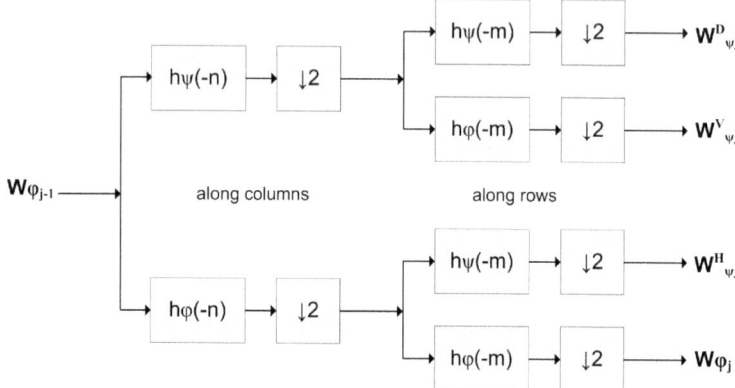

Figure A.6 : One level of decomposition for images in a *separable* 2d-wavelet system.

A.4 Summary

An important aspect in signal processing is the transformation of a class of signals into a space that brings forward certain intrinsic signal characteristics. For instance, the Fourier series representation is particularly well suited to describe periodic signals, however it is less performing for short and time-limited signals. Wavelets are extremely useful for the representation of finite support signals such as spike events in neurophysiology.

In this chapter concepts of the wavelet theory that are critical in the understanding of this thesis were communicated. It was pointed out that numerical implementations of the continuous wavelet transform can be obtained with discrete filter banks. In particular, the fast orthogonal discrete wavelet transform was introduced since it plays an important role for one possible strategy towards real-time implementations of a wavelet-based signal processing framework. A drawback of the orthogonal wavelet transform is its shift-variance. A signal and a time-shifted version thereof do not produce only shifted wavelet coefficients, but they change significantly character. This lack of shift-invariance is not very practical for feature extraction and pattern recognition. Redundant wavelet transforms were introduced. They restore the shift-invariance at the cost of higher computational complexity. Another class of wavelets, the biorthogonal wavelets, were not considered in this work and therefore not discussed in this chapter. For a detailed review of biorthogonal wavelets, one can refer to [143].

Finally, the extension of the wavelet transforms to images was briefly mentioned. The discussion was limited to *separable* scaling and wavelet functions that are built from a product of two one-dimensional functions. This *separation* enables efficient computation by feeding first the rows of an image to a one-dimensional filter pair and then followed by applying the columns from the resulting images to the second one-dimensional filter pair.

Appendix B

Abbreviations

MEA	Microelectrode Array
APS-MEA	Microelectrode Arrays based on Active Pixel Sensor Concept
R	Thermodynamic gas constant R=8.314472 $\frac{J}{K \cdot mol}$
F	Faraday constant F=96485 $\frac{C}{mol}$
CT	Computer Tomography
DOI	Diffuse Optical Imaging
PET	Positron Emission Tomography
fMRI	Functional Magnetic Resonance Imaging
NIRS	Near Infrared Spectroscopic Imaging
FET	Field Effect Transistor
CMOS	Complementary Metal Oxide Semiconductor
FPGA	Field Programmable Gate Array
DSP	Digital Signal Processor
ADC	Analog-to-Digital Converter
SNR	Signal-to-Noise Ratio
PCA	Principal Component Analysis
ICA	Independent Component Analysis
PSTH	Post Stimulus Time Histogram

CPU	Central Processing Unit
PCI	Peripheral Component Interconnect
PD	Power Density
OTA	Operational Transconductance Amplifier
LE	Logic Elements
PLL	Phase-Locked Loops
SRAM	Static Random Access Memory
LVDS	Low Voltage Differential Signaling
MDR26	Mini-D-Ribbon connector with 26 pins
LEONARDO	Frame grabber from Arvoo including RISC processor
SDRAM	Synchronous Dynamic Random Access Memory
DMA	Direct Memory Access
RAID	Redundant Array of Independent Disks
API	Application Programming Interface
SAS	Serial Attached Small Computer System Interface (SCSI)
MB	Megabyte
FIR	Finite Impulse Response
IIR	Infinite Impulse Response
DUT	Device Under Test
Pt	Platinum
TTX	Tetrodotoxin
4-AP	4-Aminopyridine
CTZ	Cyclothiazide
DIV	Days *in vitro*
WT	Wavelet Transform
DWT	Discrete Wavelet Transform
IDWT	Inverse Discrete Wavelet Transform

Abbreviations

SIDWT	Shift-Invariant Discrete Wavelet Transform
RMS	Root Mean Square
WTMM	Wavelet Transform Modulus Maxima
FP	False Positive
FN	False Negative
EZC	Energy Zero-Crossing
CWT	Contiuous Wavelet Transform
FFT	Fast Fourier Transform
INNS	Izhikevich Neural Network Simulator
CoG	Center of Gravity
VLSI	Very Large Scale Integration
CU	Computational Unit

Bibliography

[1] C. Golgi, *Opera Omnia*, Milan, Italy, 1903.

[2] A. Hodgkin and A. Huxley, "A quantitative description of membrane current and its application to conduction and excitation in nerve," *J. Physiol.*, vol. 117, pp. 500–544, 1952.

[3] E. Neher, B. Sakmann, and J. Steinbach, "The extracellular patch clamp: A method for resolving currents through individual open channels in biological membranes," *Pflügers Arch.*, vol. 375, no. 2, pp. 219–228, 1978.

[4] G. Buzsaki, "Large-scale recording of neuronal ensembles," *Nat. Neurosci.*, vol. 7, no. 5, pp. 446–451, 2004.

[5] E. Brown, R. Dass, and P. P. Mitra, "Multiple neural spike train data analysis: State-of-the-art and future challenges," *Nat. Neurosci.*, vol. 7, no. 5, pp. 456–461, 2004.

[6] T. Manuccia and L. Dobbs, "The optically switched micorelectrode array," in *Substrate-Integrated Microelectrode Arrays: Technology and Applications*, Reutlingen, 1998.

[7] L. Berdondini, T. Overstolz, N. F. De Rooij, M. Koudelka-Hep, M. Wäny, and P. Seitz, "High-density microelectrode arrays for electrophysiological activity imaging of neuronal networks," in *ICECS, the 8th IEEE International Conference on Electronics, Circuits and Systems*, vol. 3, 2001, pp. 1239–1242.

[8] M. Canepari, M. Bove, E. Maeda, M. Cappello, and A. Kawana, "Experimental analysis of neuronal dynamics in cultured cortical networks and transitions between different patterns of activity," *Biol. Cybern.*, vol. 77, pp. 153–162, 1997.

[9] Y. Jimbo, T. Tateno, and H. Robinson, "Simultaneous induction of pathway-specific potentiation and depression in networks of cortical neurons," *Biophys. J.*, vol. 76, pp. 670–678, 1999.

[10] G. Shahaf and S. Marom, "Learning in networks of cortical neurons," *J. Neurosci.*, vol. 21, no. 22, pp. 8782–8788, 2001.

[11] J. van Pelt, P. Wolters, M. Corner, W. Rutten, and G. Ramakers, "Long-term characterization of firing dynamics of spontaneous bursts in cultured neural networks," *IEEE Trans. Biomed. Eng.*, vol. 51, no. 11, pp. 2051–2062, 2004.

[12] L. Stoppini, P. Buchs, and D. Muller, "A simple method for organotypic cultures of nervous tissue," *J. Neurosci. Meth.*, vol. 37, pp. 173–182, 1991.

[13] E. Schneidman, W. Bialek, and M. Berry, "Synergy, redundancy and independence in population codes," *J. Neurosci.*, vol. 23, no. 37, pp. 11 539–11 553, 2003.

[14] E. Schneidman, M. Berry, R. Segev, and W. Bialek, "Weak pairwise correlations imply strongly correlated network states in a neural population," *Nature*, vol. 440, 2006.

[15] M. Laubach, J. Wessberg, and M. Nicolelis, "Cortical ensemble activity increasingly predicts behaviour outcomes during learning of a motor task," *Nature*, vol. 405, pp. 567–571, 2000.

[16] T. DeMarse, D. Wagenaar, A. Blau, and S. Potter, "The neurally controlled animat: Biological brains acting with simulated bodies," *Auton. Robot.*, vol. 11, no. 3, pp. 305–310, 2001.

[17] S. Potter, "Distributed processing in cultured neuronal networks," in *Progress in Brain Research: Advances in Neural Population Coding*, M. Nicolelis, Ed., Amsterdam, 2001, vol. 130, pp. 49–62.

[18] S. Potter and T. DeMarse, "A new approach to neural cell culture for long-term studies," *J. Neurosci. Meth.*, vol. 110, pp. 17–24, 2001.

[19] A. Stett, U. Egert, E. Guenther, F. Hofmann, T. Meyer, W. Nisch, and H. Haemmerle, "Biological application of microelectrode arrays in drug discovery and basic research," *Anal. Bioanal. Chem.*, vol. 377, pp. 486–495, 2003.

[20] G. W. Gross and B. Rhoades, "The use of neuronal networks on multielectrode arrays as biosensors," *Biosensors & Bioelectronics*, vol. 10, pp. 553–567, 1995.

[21] I. Antonov, I. Antonova, E. Kandel, and R. Hawkins, "Activity-dependent presynaptic facilitation and hebbian LTP are both required and interact during classical conditioning in aplysia," *Neuron*, vol. 37, pp. 135–147, 2003.

[22] E. Keefer, A. Gramowski, D. Stenger, J. Pancrazio, and G. Gross, "Characterization of acute neurotoxic effects of trimethylolpropane phospate

via neuronal network biosensors," *Biosensors & Bioelectronics*, vol. 16, pp. 513–525, 2001.

[23] K. Shimono, M. Baudry, V. Panchenko, and M. Taketani, "Chronic multichannel recordings from organotypic hippocampal slice cultures: Protection from excitotoxic effects of Nmda by non-competitive Nmda antagonists," *J. Neurosci. Meth.*, vol. 120, pp. 193–202, 2002.

[24] J. Wessberg, C. Stambaugh, J. Kralik, P. Beck, M. Laubach, J. Chapin, J. Kim, J. Biggs, M. Srinivasan, and M. Nicolelis, "Real-time prediction of hand trajectory by ensembles of cortical neurons in primates," *Nature*, vol. 408, pp. 361–365, 2000.

[25] M. Serruya, N. Hatsopoulos, L. Paninski, M. Fellows, and J. Donoghue, "Instant neural control of a movement signal," *Nature*, vol. 416, pp. 141–142, 2002.

[26] J. Donoghue, "Connecting cortex to machines: Recent advances in brain interfaces," *Nat. Neurosci.*, vol. 5, pp. 1085–1088, 2002.

[27] F. Mussa-Ivaldi and L. Miller, "Brain-machine interfaces: Computational demands and clinical needs meet basic neuroscience," *Trends in Neuroscience*, vol. 26, no. 6, pp. 329–332, 2003.

[28] H. Berry and O. Temam, "Modeling self-developing biological neural networks," *Neurocomputing*, vol. 70, pp. 2723–2734, 2007.

[29] A. Destexhe, Z. Mainen, and T. Sejnowski, "Synthesis of models for excitable membranes, synaptic transmission and neuromodulation using a common kinetic formalism," *J. Comput. Neurosci.*, vol. 1, no. 195-230, 1994.

[30] B. Ermentrout, "Neural networks as spatio-temporal pattern-forming systems," *Rep. Prog. Phys.*, vol. 61, pp. 353–430, 1998.

[31] D. Golomb, A. Shedmi, R. Curtu, and B. Ermentrout, "Persistent synchronized bursting activity in cortical tissues with low magnesium concentration: A modeling study," *J. Neurophysiol.*, vol. 95, pp. 1049–1067, 2006.

[32] D. Golomb, C. Yue, and Y. Yaari, "Contribution of persistent na+ current and m-type k+ current to somatic bursting in CA1 pyramidal cells: Combined experimental and modeling study," *J. Neurophysiol.*, vol. 96, pp. 1912–1926, 2007.

[33] A. Bard and L. Faulkner, *Electrochemical Methods*. New York: John Wiley & Sons, Inc., 1980.

[34] M. Bear, B. Connors, and M. Paradiso, *Neuroscience - Exploring the Brain*. Baltimore: Williams & Wilkins, 1996.

[35] J. Nicholls, A. Martin, B. Wallace, and P. Fuchs, *From Neuron to Brain*, 4th ed. Sunderland: Sinauer, 2001.

[36] M. Nicolelis, A. Ghazanfar, C. Stambaugh, L. Oliveira, M. Laubach, J. Chapin, R. Nelson, and J. Kaas, "Simultaneous encoding of tactile information by three primate cortical areas," *Nat. Neurosci.*, vol. 1, no. 7, pp. 621–630, 1998.

[37] J. Puchalla, E. Schneidman, R. Harris, and M. Berry, "Redundancy in the population code of the retina," *Neuron*, vol. 46, pp. 493–504, 2005.

[38] G. Gross, "Simultaneous single unit recording in vitro with a photoetched laser deinsulated gold multimicroelectrode surface," *IEEE Trans. Biomed. Eng.*, vol. 26, no. 5, pp. 273–279, 1979.

[39] J. Pine, "Recording action potentials from cultured neurons with extracellular microcircuit electrodes," *J. Neurosci. Meth.*, vol. 2, pp. 19–31, 1980.

[40] D. Robinson, "The electrical properties of metal microelectrodes," *Proc. IEEE*, vol. 56, no. 6, pp. 1065–1071, 1968.

[41] M. Grattarola and S. Martinioia, "Modeling the neuron-microtransducer junction: From extracellular to patch recording," *IEEE Trans. Biomed. Eng.*, vol. 40, no. 1, p. 35, 1993.

[42] D. Borkholder, "Cell based biosensors using microeletrodes," Ph.D. dissertation, University of California, 1998.

[43] J. Buitenweg, W. Rutten, E. Marani, S. Polman, and J. Ursum, "Extracellular detection of active membrane currents in the neuron-electrode interface," *J. Neurosci. Meth.*, vol. 115, pp. 211–221, 2002.

[44] S. Martinoia, P. Massobrio, M. Bove, and G. Massobrio, "Cultured neurons coupled to microelectrode arrays: Circuit models, simulations and experimental data," *IEEE Trans. Biomed. Eng.*, vol. 51, no. 5, pp. 859–864, 2004.

[45] M. Bove, S. Martinoia, M. Grattarola, and D. Ricci, "The neuron-transistor junction: Linking equivalent circuit models to microscopic descriptions," *Thin Solid Films*, vol. 284-285, pp. 772–775, 1996.

[46] J. Buitenweg, "Electrical behaviour of the neuron-electrode interface," Ph.D. dissertation, Twente University Press, 2001.

[47] L. Berdondini, "Nano- and microfabricated interfaces for in-vitro electrophysiology," Ph.D. dissertation, Neuchâtel, 2003.

[48] J. Buitenweg, W. Rutten, W. Willems, and J. van Nieuwkasteele, "Measurement of sealing resistance of cell-electrode interfaces in neuronal cultures using impedance spectroscopy," *Med. Biol. Eng. Comput.*, vol. 36, no. 5, pp. 630–637, 1998.

[49] C. Thomas, P. Springer, G. Loeb, Y. Berwald-Netter, and L. Okun, "A miniature microelectrode array to monitor the bioelectric activity of cultured cells," *Exp. Cell Res.*, vol. 74, pp. 61–66, 1972.

[50] G. W. Gross, A. Williams, and J. Lucas, "Recording of spontaneous activity with photoetched microelectrode surfaces from mouse spinal neurons in culture," *J. Neurosci. Meth.*, vol. 5, pp. 13–22, 1982.

[51] K. Mathieson, S. Kachiguine, C. Adams, W. Cunningham, D. CGunning, V. O'Shea, K. Smith, E. Chichilnisky, A. M. Litke, A. Sher, and M. Rahman, "Large-area microelectrode arrays for recording of neural signals," *IEEE Trans. Nucl. Sci.*, vol. 51, no. 5, pp. 2027–2031, 2004.

[52] B. DeBusschere and G. Kovacs, "Portable cell-based biosensor system using integrated CMOS cell-cartridges," *Biosensors & Bioelectronics*, vol. 16, pp. 543–556, 2001.

[53] R. Harrison and C. Charles, "A low-power low-noise CMOS amplifier for neural recording applications," *IEEE J. Solid-State Circ.*, vol. 38, no. 6, pp. 958–965, 2003.

[54] W. Dabrowski, P. Grybos, and A. M. Litke, "A low noise multichannel integrated circuit for recording neuronal signals using microelectrode arrays," *Biosensors & Bioelectronics*, vol. 19, pp. 749–761, 2004.

[55] P. Mohseni and K. Najafi, "A fully integrated neural recording amplifier with DC input stabilization," *IEEE Trans. Biomed. Eng.*, vol. 51, no. 5, pp. 832–837, 2004.

[56] K. Wise, J. Angell, and A. Starr, "An integrated circuit approach to extracellular microelectrodes," *IEEE Trans. Biomed. Eng.*, vol. 17, pp. 238–246, 1970.

[57] K. Najafi and K. Wise, "An implantable multielectrode array with on-chip signal processing," *IEEE J. Solid-State Circ.*, vol. sc-21, no. 6, pp. 1035–1044, 1986.

[58] F. Heer, W. Franks, A. Blau, S. Taschini, C. Ziegler, A. Hierlemann, and H. Baltes, "CMOS microelectrode array for the monitoring of electrogenic cells," *Biosensors & Bioelectronics*, 2004.

[59] F. Heer, S. Hafizovic, W. Franks, A. Blau, C. Ziegler, and A. Hierlemann, "CMOS micorelectrode array for bidirectional interaction with neuronal networks," *IEEE J. Solid-State Circ.*, vol. 41, no. 7, pp. 1620–1629, 2006.

[60] O. Billoint, J.-P. Rostaing, G. Charvet, and B. Yvert, "A 64-channel ASIC for in-vitro simultaneous recording and stimulation of neurons using microelectrode arrays," in *29th Int'l Conf. of IEEE EMBS*, Lyon, 2007, pp. 6069–6072.

[61] B. Eversmann, M. Jenkner, F. Hofmann, C. Paulus, R. Brederlow, B. Holzapfl, P. Fromherz, M. Merz, M. Brenner, M. Schreiter, R. Gabl, K. Plehnert, M. Steinhauser, G. Eckstein, D. Schmitt-Landsiedel, and R. Thewes, "A 128x128 CMOS biosensor array for extracellular recording of neural activity," *IEEE J. Solid-State Circ.*, vol. 38, no. 12, pp. 2306–2317, 2003.

[62] P. Fromherz, A. Offenhusser, T. Vetter, and J. Weis, "A neuron-silicon junction: A retzius cell of the leech on an insulated-gate field-effect transistor," *Science, New Series*, vol. 252, no. 5010, pp. 1290–1293, 1991.

[63] M. Hutzler, A. Lambacher, B. Eversmann, M. Jenkner, R. Thewes, and P. Fromherz, "High-resolution multitransistor array recording of electrical field potentials in cultured brain slices," *J. Neurophysiol.*, vol. 96, pp. 1638–1645, 2006.

[64] S. Hafizovic, F. Heer, W. Franks, F. Greve, A. Blau, C. Ziegler, and A. Hierlemann, "Cmos bidirectional electrode array for electrogenic cells," in *MEMS*, Istanbul, 2006, pp. 4–7.

[65] T. Borghi, R. Gusmeroli, A. Spinelli, and G. Baranauskas, "A simple method for efficient spike detection in multiunit recordings," *J. Neurosci. Meth.*, vol. 163, pp. 176–180, 2007.

[66] K. Oweiss, A. Mason, Y. Suhail, A. Kamboh, and K. Thomson, "A scalable wavelet transform VLSI architecture for real-time signal processing in high-density intra-cortical implants," *IEEE Trans. Circuits. Syst. - I*, vol. 54, no. 6, pp. 1266–1278, 2007.

[67] K. S. Guillory and R. A. Normann, "A 100-channel system for real time detection and storage of extracellular spike waveforms," *J. Neurosci. Meth.*, vol. 91, pp. 21–29, 1999.

[68] A. Folgers, F. Msch, T. Malina, and U. Hofmann, "Realtime bioelectrical data acquisition and processing from 128 channels utilizing the wavelet-transformation," *Neurocomputing*, vol. 52-54, pp. 247–254, 2003.

[69] J. Proakis and D. Manolakis, *Digital Signal Processing - Principles, Algorithms and Applications*, 3rd ed. New Jersey: Prentice Hall, 1996.

[70] D. Manolakis, V. Ingle, and S. Kogon, *Statistical and Adaptive Signal Processing*. Boston: Artech House, 2005.

[71] S. N. Gozani and J. P. Miller, "Optimal discrimination and classification of neuronal action potential waveforms from multiunit, multchannel recordings using software-base linear filters," *IEEE Trans. Biomed. Eng.*, vol. 41, no. 4, p. 358, 1994.

[72] A. M. Rao and D. L. Jones, "A denoising approach to multisensor signal estimation," *IEEE Trans. Signal Process.*, vol. 48, no. 5, p. 1225, 2000.

[73] K. G. Oweiss and D. J. Anderson, "Noise reduction in multichannel neural recordings using a new array wavelet denoising algorithm," *Neurocomputing*, vol. 38-40, pp. 1687–1693, 2001.

[74] P. Comon, "Independent component analysis, a new concept ?" *Signal Process.*, vol. 36, pp. 287–314, 1994.

[75] A. Bell and T. Sejnowski, "An information-maximization approach to blind separation and blind deconvolution," *Neural Comput.*, vol. 7, pp. 1129–1159, 1995.

[76] J.-F. Cardoso, "Blind signal separation: Statistical principles," *Proc. IEEE*, vol. 86, no. 10, pp. 2009–2025, 1998.

[77] A. Hyvrinen, J. Karhunen, and E. Oja, *Independent Component Analysis*. John Wiley & Sons, 2001.

[78] H. Krim and M. Viberg, "Two decades of array signal processing research," *IEEE Signal Proc. Mag.*, pp. 67–94, 1996.

[79] M. Laubach, M. Shuler, and M. Nicolelis, "Independent component analyses for quantifying neuronal ensemble interactions," *J. Neurosci. Meth.*, vol. 94, pp. 141–154, 1999.

[80] J. Chapin and M. Nicolelis, "Principal component analysis of neuronal ensemble activity reveals multidimensional somatosensory representations," *J. Neurosci. Meth.*, vol. 94, no. 1, pp. 121–140, 1999.

[81] K. G. Oweiss and D. J. Anderson, "A unified framework for advancing array signal processing technology of multichannel microprobe neural recordings devices," in *2nd annual International IEEE-EMBS Special Topic Conference on Microtechnologies in Medicine & Biology*, Madison, Wisconsin, 2002, pp. 245–250.

[82] D. H. Perkel, G. L. Gerstein, and G. P. Moore, "Neuronal spike trains and stochastic point processes i.the single spike train," *Biophys. J.*, vol. 7, no. 4, pp. 391–418, 1967.

[83] M. S. Lewicki, "A review of methods for spike sorting: The detection and classification of neural action potentials," *Network: Comput. Neural. Syst.*, vol. 9, p. R53, 1998.

[84] I. Obeid and P. Wolf, "Evaluation of spike-detection algorithms for a brain-machine interface application," *Trans. Biomed. Eng.*, vol. 51, no. 6, pp. 905–911, 2004.

[85] H. Nakatani, T. Watanabe, and N. Hoshimiya, "Detection of nerve action potentials under low signal-to-noise ratio condition," *IEEE Trans. Biomed. Eng.*, vol. 48, no. 8, pp. 845–849, 2001.

[86] R. Quian Quiroga, O. Sakowitz, E. Basar, and M. Schrmann, "Wavelet transform in the analysis of the frequency composition of evoked potentials," *Brain Res. Protoc.*, vol. 8, pp. 16–24, 2001.

[87] E. Hulata, R. Segev, and E. Ben-Jacob, "A method for spike sorting and detection based on wavelets packets and shannon's mutual information," *J. Neurosci. Meth.*, vol. 117, pp. 1–12, 2002.

[88] K. Kim and S. Kim, "A wavelet-based method for action potential detection from extracellular neural signal recording with low signal-to-noise ratio," *IEEE Trans. Biomed. Eng.*, vol. 50, no. 8, pp. 999–1011, 2003.

[89] Z. Nenadic and J. Burdick, "Spike detection using the continuous wavelet transform," *IEEE Trans. Biomed. Eng.*, vol. 52, no. 1, pp. 74–87, 2005.

[90] M. Lewicki, "Bayesian modeling and classification of neural signals," *Neural Comput.*, vol. 6, no. 5, pp. 1005–1030, 1994.

[91] B. Wheeler and W. Heetderks, "A comparison of techniques for classification of multiple neural signals," *IEEE Trans. Biomed. Eng.*, vol. 29, no. 12, pp. 752–759, 1982.

[92] J. C. Letelier and P. P. Weber, "Spike sorting based on discrete wavelet transform coefficients," *J Neurosci. Meth.*, vol. 101, pp. 93–106, 2000.

[93] A. Maccione, "An innovative approach to neuronal network dynamics studies with new high density micro electrode arrays," Ph.D. dissertation, University of Genova, 2008.

[94] V. Makarov, F. Panetsos, and O. De Feo, "A method for determining neural connectivity and inferring the underlying network dynamics using

extracellular spike recordings," *J. Neurosci. Meth.*, vol. 144, pp. 265–279, 2005.

[95] I. Nemenman, W. Bialek, and R. de Ruyter van Steveninck, "Entropy and information in neural spike trains: Progress on the sampling problem," *Phys. Rev. E*, vol. 69, 2004.

[96] P. Latham and S. Nirenberg, "Synergy, redundancy and independence in population codes, revisited," *J. Neurosci.*, vol. 25, no. 21, pp. 5195–5206, 2005.

[97] C. Santa Maria, "Optimization of cell culture procedures for growing neural networks on microelectrode arrays," Ph.D. dissertation, University of North Texas, 2007.

[98] K. Kitano and T. Fukai, "Variability v.s. synchronicity of neuronal activity in local cortical network models with different wiring topologies," *J. Comput. Neurosci.*, vol. 23, pp. 237–250, 2007.

[99] L. Berdondini, P. D. Van der Wal, O. Guenat, N. F. de Rooij, M. Koudelka-Hep, P. Seitz, R. Kaufmann, P. Metzler, N. Blanc, and S. Rohr, "High-density electrode array for imaging in vitro electrophysiological activity," *Biosensors & Bioelectronics*, vol. 21, no. 1, pp. 167–174, 2005.

[100] B. Razavi, *Design of Analog CMOS Integrated Circuits*. New York: Mc-Graw Hill, 2001.

[101] U. Frey, C. Sanchez-Bustamante, T. Ugniwenko, F. Heer, J. Sedify, S. Hafizovic, B. Roscic, M. Fussenegger, A. Blau, U. Egert, and A. Hierlemann, "Cell recordings with a CMOS high-density microelectrode array," in *Proc. of the 29th International IEEE EMBS Conf.*, Lyon, 2007, pp. 167–170.

[102] D. A. Wagenaar, J. Pine, and S. M. Potter, "An extremely rich repertoire of bursting patterns during the development of cortical cultures," *BMC Neurosci.*, vol. 7, no. 11, 2006.

[103] L. Berdondini, P. Massobrio, M. Chiappalone, M. Tedesco, K. Imfeld, A. Maccione, M. Koudelka-Hep, and S. Martinoia, "Extracellular recordings from locally dense microelectrode arrays coupled to dissociated cortical cultures," *J. Neurosci. Meth.*, vol. doi:10.1016/j.jneumeth.2008.10.032, 2008.

[104] S. Mendis, S. E. Kemeny, and E. R. Fossum, "CMOS active pixel image sensor," *IEEE Trans. Electron Dev.*, vol. 41, no. 3, pp. 452–453, 1994.

[105] R. Nixon, S. E. Kemeny, B. Pain, C. Staller, and E. R. Fossum, "256x256 CMOS active pixel sensor camera-on-a-chip," *IEEE J. Solid-State Circ.*, vol. 31, no. 12, pp. 2046–2050, 1996.

[106] A. Krymski and N. Tu, "A 9 V/Lux-s 5000 frames/s 512x512 CMOS sensor," *IEEE Trans. Electron Dev.*, vol. 50, no. 1, pp. 136–143, 2003.

[107] S. Kleinfelder, Y. Chen, K. Kwiatkowski, and A. Shah, "High-speed CMOS image sensor circuits with in-situ frame storage," *IEEE Trans. Nucl. Sci.*, vol. 51, no. 4, pp. 1648–1656, 2004.

[108] K. G. Oweiss, "A systems approach for data compression and latency reduction in cortically controlled brain machine interfaces," *IEEE Trans. Biomed. Eng.*, vol. 53, no. 7, pp. 1364–1377, 2006.

[109] P. Dayan and L. Abbott, *Theoretical Neuroscience - Computational and Mathematical Modeling of Neural Systems*. Cambridge: The MIT Press, 2001.

[110] K. Kim and S. Kim, "Noise performance design of CMOS preamplifier for the active semiconductor neural probe," *IEEE Trans. Biomed. Eng.*, vol. 47, no. 8, pp. 1097–1105, 2000.

[111] B. Razavi, *Rf Microelectronics*. Upper Saddle River, NJ: Prentice Hall PTR, 1998.

[112] M. Chiappalone, M. Bove, A. Vato, M. Tedesco, and S. Martinoia, "Dissociated cortical networks show spontaneously correlated activity patterns during in vitro development," *Brain Res.*, vol. 1093, no. 1, pp. 41–53, 2006.

[113] D. Joanes and C. Gill, "Comparing measures of sample skewness and kurtosis," *The Statistician*, vol. 47, pp. 183–189, 1998.

[114] J. Qi, Y. Wang, M. Jiang, P. Warren, and G. Chen, "Cyclothiazide induces robust epileptiform activity in rat hippocampal neurons both in vitro and in vivo," *J. Physiol.*, vol. 571, pp. 605–618, 2006.

[115] K. Imfeld, S. Neukom, A. Maccione, Y. Bornat, S. Martinoia, P.-A. Farine, M. Koudelka-Hep, and L. Berdondini, "Large-scale, high-resolution data acquisition system for extracellular recording of electrophysiological activity," *IEEE Trans. Biomed. Eng.*, vol. 55, no. 8, pp. 2064–2073, 2008.

[116] K. G. Oweiss and D. J. Anderson, "Spike sorting: A novel shift an amplitude invariant technique," *Neurocomputing*, vol. 44-46, p. 1133, 2002.

[117] M. Laubach, "Wavelet-based processing of neuronal spike trains prior to discriminant analysis," *J. Neurosci. Meth.*, vol. 134, pp. 159–168, 2004.

[118] I. Daubechies, *Ten Lectures on Wavelets*. Philadelphia, PA: SIAM, 1992.

[119] G. Beylkin, R. Coifman, and V. Rokhlin, "Fast wavelet transforms and numerical algorithms i," *Comm. Pure Appl. Math.*, vol. 44, pp. 141–183, 1991.

[120] D. H. Perkel, G. L. Gerstein, and G. P. Moore, "Neuronal spike trains and stochastic point processes ii. simultaneous spike trains," *Biophys. J.*, vol. 7, no. 4, pp. 419–440, 1967.

[121] G. P. Moore, D. H. Perkel, and J. Segundo, "Statistical analysis and functional interpretation of neural spike data," *Annu. Rev. Physiol.*, vol. 28, pp. 493–522, 1966.

[122] M. Abeles and M. Goldstein, "Multispike train analysis," *Proc. IEEE*, vol. 65, no. 5, pp. 762–776, 1977.

[123] M. Stecker, "Effect of neural connectivity on autocovariance and cross covariance estimates," *BioMedical Engineering Online*, vol. 6, no. 3, 2007.

[124] R. Chandra and L. Optican, "Detection, classification and superposition resolution of action potentials in multiunit single-channel recordings by an on-line real-time neural network," *IEEE Trans. Biomed. Eng.*, vol. 44, pp. 403–412, 1997.

[125] P. Thakur, H. Lu, S. Hsiao, and K. Johnson, "Automated optimal detection and classification of neural action potentials in extra-cellular recordings," *J. Neurosci. Meth.*, vol. 162, no. 1-2, pp. 364–376, 2007.

[126] B. Wheeler, "Automatic discrimination of single units," in *Methods for Neural Ensemble Recordings*, M. Nicolelis, Ed. New York: CRC Press, 1999, pp. 61–77.

[127] S. Shoham, M. Fellows, and R. Normann, "Robust, automatic spike sorting using mixtures of multivariate t-distributions," *J. Neurosci. Meth.*, vol. 127, pp. 111–122, 2003.

[128] L. Kaufman and P. Rousseeuw, *Finding Groups in Data: An Introduction to Cluster Analysis*. New York: John Wiley & Sons, 1990.

[129] D. Adamos, E. Kosmidis, and G. Theophilidis, "Performance evaluation of PCA-based spike sorting algorithms," *Comput. Methods Programs Biomed.*, vol. 91, no. 3, pp. 232–244, 2008.

[130] A. Maccione, M. Gandolfo, P. Massobrio, A. Novellino, S. Martinoia, and M. Chiappalone, "A novel algorithm for precise identification of spikes in extracellularly recorded neuronal signals," *J. Neurosci. Meth.*, vol. doi: 10.1016/j.jneumeth.2008.09.026.

[131] M. Maher, J. Pine, J. Wright, and Y.-C. Tai, "The neurochip: A new multielectrode device for stimulating and recording from cultured neurons," *J. Neurosci. Meth.*, vol. 87, pp. 45–56, 1999.

[132] K. Wise, D. J. Anderson, J. Hetke, D. Kipke, and K. Najafi, "Wireless implantable microsystems: High-density electronic interfaces to the nervous system," *Proc. IEEE*, vol. 92, no. 1, pp. 76–97, 2004.

[133] M. Salganicoff, M. Sarna, L. Sax, and G. Gerstein, "Unsupervised waveform classification for multineuron recordings: A real-time software based system, i: Algorithms and implementation," *J. Neurosci. Meth.*, vol. 25, no. 181-187, 1988.

[134] U. Rutishauser, E. Schuman, and A. Mamelak, "Online detection and sorting of extracellularly recorded action potentials in human medial temporal lobe recordings, in vivo," *J. Neurosci. Meth.*, vol. 154, no. 1-2, pp. 204–224, 2006.

[135] D. L. Donoho, "De-noising by soft-thresholding," *IEEE Trans. Inform. Theory*, vol. 41, no. 3, pp. 613–627, 1995.

[136] M. Lang, H. Guo, J. Odegard, C. Burrus, and R. Wells, "Noise reduction using an undecimated discrete wavelet transform," *IEEE Signal Proc. Let.*, vol. 3, no. 1, pp. 10–12, 1996.

[137] R. Quian Quiroga, Z. Nadasdy, and Y. Ben-Shaul, "Unsupervised spike detection and sorting with wavelets and superparamagnetic clustering," *Neural Comput.*, vol. 16, pp. 1661–1687, 2004.

[138] G. Zouridakis and D. Tam, "Multi-unit spike discrimination using wavelet transforms," *Comput. Biol. Med*, vol. 27, no. 1, pp. 9–18, 1997.

[139] S. Grace Chang, B. Yu, and M. Vetterli, "Adaptive wavelet thresholding for image denoising and compression," *IEEE Trans. Image Process.*, vol. 9, no. 9, pp. 1532–1546, 2000.

[140] S. Mallat and W. L. Hwang, "Singularity detection and processing with wavelets," *IEEE Trans. Inform. Theory*, vol. 38, no. 2, pp. 617–643, 1992.

[141] H. Szu, Y. Sheng, and J. Chen, "Wavelet transform as a bank of the matched filters," *Appl. Opt.*, vol. 31, no. 17, pp. 3267–3277, 1992.

[142] M. Unser and A. Aldroubi, "A review of wavelets in biomedical applications," *Proc. IEEE*, vol. 84, no. 4, pp. 626–638, 1996.

[143] S. Mallat, *A Wavelet Tour of Signal Processing*. New York: Academic Press, 2001.

[144] C. Burrus, R. Gopinath, and H. Guo, *Introduction to Wavelets and Wavelet Transforms.* Prentice Hall, 1998.

[145] S. Mallat, "Dyadic wavelets energy zero-crossings," Department of Computer & Information Science, University of Pennsylvania, Tech. Rep., 1988.

[146] M. Shensa, "The discrete wavelet transform: Wedding the a trous and mallat algorithms," *IEEE Trans. Signal Proces.*, vol. 40, no. 10, pp. 2464–2482, 1992.

[147] S. G. Mallat, "A theory for multiresolution signal decomposition: The wavelet representation," *IEEE Trans. Pattern Anal.*, vol. 11, no. 7, pp. 674–693, 1989.

[148] O. Rioul and P. Duhamel, "Fast algorithms for discrete and continuous wavelet transforms," *IEEE Trans. Inform. Theory.*, vol. 38, no. 2, pp. 569–586, 1992.

[149] M. Vetterli, "Running FIR and IIR filtering using multirate filter banks," *IEEE Trans. Signal Process.*, vol. 36, pp. 730–738, 1988.

[150] H. Krim, D. Tucker, S. Mallat, and D. Donoho, "On denoising and best signal representation," *IEEE Trans. Inform. Theory*, vol. 45, no. 7, pp. 2225–2238, 1999.

[151] I. Selesnick, R. Baraniuk, and N. Kingsbury, "The dual-tree complex wavelet transform," *Signal Processing Magazine*, pp. 124–152, 2005.

[152] K. Oweiss, "An information theoretic approach for optimizing the analysis of multi-electrode neuronal recordings," in *26th Annu. Inter'l Conf. IEEE Eng. Med. Biol. Soc.*, San Francisco, 2004, pp. 4520–4523.

[153] D. Johnson and D. Dudgeon, *Array Signal Processing.* Prentice Hall, 1993.

[154] B. Van Veen and K. Buckley, "Beamforming: A versatile approach to spatial filtering," *IEEE ASSP Magazine*, pp. 4–24, 1988.

[155] W. Roberts and D. Hartline, "Separation of multi-unit nerve impulse trains by a multi-channel linear filter algorithm," *Brain Res.*, vol. 94, pp. 141–149, 1975.

[156] E. Schmidt, "Computer separation of multi-unit neuroelectric data: A review," *J. Neurosci. Meth.*, vol. 12, pp. 95–111, 1984.

[157] E. Balster, Y. Zheng, and R. Ewing, "Feature-based wavelet shrinkage algorithm for image denoising," *IEEE Trans. Image Process.*, vol. 14, no. 12, pp. 2024–2039, 2005.

[158] J. L. Starck, E. J. Cands, and D. L. Donoho, "The curvelet transform for image denoising," *IEEE Trans. Image Process.*, vol. 11, no. 6, pp. 670–684, 2002, curvelet; Discrete wavelet transform; FFT; Filtering; FWT; Radon transform; Ridgelets; Thresholding rules; Wavelets Date of Input: 07.06.2004 Priority: Normal.

[159] L. Bretzner, "Multi-scale feature tracking and motion estimation," Ph.D. dissertation, University of Stockholm, 1999.

[160] M. Unser, "Texture classification and segmentation using wavelet frames," *IEEE Trans. Image Process.*, vol. 4, no. 11, pp. 1549–1560, 1995.

[161] Y. Xu, J. Weaver, D. Healy, and J. Lu, "Wavelet transform domain filters: A spatially selective noise filtration technique," *IEEE Trans. Image Process.*, vol. 3, no. 6, pp. 747–758, 1994.

[162] N. Carnevale and M. Hines, *The Neuron Book*. Cambridge, UK: Cambridge University Press, 2006.

[163] J. Modolo, A. Garenne, J. Henry, and A. Beuter, "Development and validation of a neural population model based on the dynamics of a discontinuous membrane potential neuron model," *J. Integr. Neurosci.*, vol. 6, no. 4, pp. 625–655, 2007.

[164] E. Izhikevich, "Simple model of spiking neurons," *IEEE Trans. Neural Networks*, vol. 14, pp. 1569–1572, 2003.

[165] ——, "Which model to use for cortical spiking neurons?" *IEEE Trans. Neural Networks*, vol. 15, pp. 1063–1070, 2004.

[166] E. Izhikevich, J. Gally, and G. Edelman, "Spike-timing dynamics of neuronal groups," *Cereb. Cortex*, vol. 14, pp. 933–944, 2004.

[167] R. Zucker, "Short-term synaptic plasticity," *Annu. Rev. Neurosci.*, vol. 12, pp. 13–31, 1989.

[168] D. Henze, Z. Borhegyi, J. Csicsvari, A. Mamiya, K. Harris, and G. Buzsaki, "Intracellular features predicted by extracellular recordings in the hippocampus in vivo," *J. Neurophysiol.*, vol. 84, no. 1, pp. 390–400, 2000.

[169] g. Brown, S. Yamada, and T. Sejnowski, "Independent component analysis at the neural cocktail party," *Trends in Neuroscience*, vol. 24, no. 1, pp. 54–63, 2001.

[170] A. Buades, B. Coll, and J. Morel, "A review of image denoising algorithms, with a new one," *Multiscale Model. Simul.*, vol. 4, no. 2, pp. 490–530, 2005.

[171] S. Mallat and S. Zhong, "Characterization of signals from multiscale edges," *IEEE Trans. Pattern Anal.*, vol. 14, no. 7, pp. 710–732, 1992.

[172] L. Zhang and P. Bao, "Denoising by spatial correlation thresholding," *IEEE Trans. Circ. Syst. Vid.*, vol. 13, no. 6, pp. 535–538, 2003.

[173] A. Pizurica, V. Zlokolica, and W. Philips, "Combined wavelet domain and temporal video denoising," in *Proc. of the IEEE International Conf. on Advanced Video and Signal Based Surveillance*, Miami, 2003, pp. 334–341.

[174] I. Selesnick and K.-L. Li, "Video denoising using 2d and 3d dual-tree complex wavelet transforms," in *Proc. Wavelet Applications Signal Image Processing X (SPIE 5207)*, San Diego, 2003, pp. 607–618.

[175] V. Zlokolica, A. Pizurica, and W. Philips, "Wavelet-domain video denoising based on reliability measures," *IEEE Trans. Circ. Syst. Vid.*, vol. 16, no. 8, pp. 993–1007, 2006.

[176] A. Dima, "Computer aided image segmentation and graph construction of nerve cells from 3d confocal microscopy scans," Ph.D. dissertation, Technical University, 2002.

[177] D. Donoho and I. Johnstone, "Adapting to unknown smoothness via wavelet shrinkage," *J. Am. Stat. Assoc.*, vol. 90, no. 432, pp. 1200–1224, 1995.

[178] F. Shi and I. Selesnick, "Video denoising using oriented complex wavelet transform," in *IEEE Int'l Conf. on Acoustics, Speech and Signal Processing*, vol. 2, 2004, pp. 949–952.

[179] E. Maeda, H. Robinson, and A. Kawana, "The mechanisms of generation and propagation of synchronized bursting in developing networks of cortical neurons," *J. Neurosci.*, vol. 15, no. 10, pp. 6834–6845, 1995.

[180] J. Beggs and D. Plenz, "Neuronal avalanches are diverse and precise activity patterns that are stable for many hours in cortical slice cultures," *J. Neurosci.*, vol. 24, no. 22, pp. 5216–5229, 2004.

[181] D. Eytan and S. Marom, "Dynamics and effective topology underlying synchronization in networks of cortical neurons," *J. Neurosci.*, vol. 26, no. 33, pp. 8465–8476, 2006.

[182] R. Madhavan, Z. Chao, and S. Potter, "Plasticity of recurring spatiotemporal activity patterns in cortical networks," *Phys. Biol.*, vol. 4, pp. 181–193, 2007.

[183] M. Antonini, M. Barlaud, P. Mathieu, and I. Daubechies, "Image coding using wavelet transform," *IEEE Trans. Image Process.*, vol. 1, no. 2, pp. 205–220, 1992.

[184] Z. Chao, D. Bakkum, D. Wagenaar, and S. Potter, "Effects of random external background stimulation on network synaptic stability after tetanization," *Neuroinformatics*, vol. 3, no. 263-280, 2005.

[185] J. Stegenga, J. Le Feber, E. Marani, and W. Rutten, "Analysis of cultured neuronal networks using intraburst firing characteristics," *IEEE Trans. Biomed. Eng.*, vol. 55, no. 4, pp. 1382–1390, 2008.

[186] G. Knowles, "VLSI architecture for the discrete wavelet transform," *Electron. Lett.*, vol. 26, no. 15, pp. 1184–1185, 1990.

[187] K. Parhi and T. Nishitani, "VLSI architectures for discrete wavelet transforms," *IEEE Trans. VLSI Syst.*, vol. 1, no. 2, pp. 191–202, 1993.

[188] P. Wu and L.-G. Chen, "An efficient architecture for two-dimensional discrete wavelet transform," *IEEE Trans. Circ. Syst. Vid.*, vol. 11, pp. 536–544, 2001.

[189] M. Vishwanath, R. Owens, and M. Irwin, "VLSI architectures for the discrete wavelet transform," *IEEE. Trans. Circ. Syst. - II: Anal. Dig. Signal Process.*, vol. 42, no. 5, pp. 305–316, 1995.

[190] W. Sweldens, "The lifting scheme: A custom-design construction of biorthogonal wavelets," *Appl. Comput. Harmon. Anal.*, vol. 3, no. 15, pp. 186–200, 1996.

[191] I. Daubechies and W. Sweldens, "Factoring wavelet transforms into lifting steps," *J. Fourier Anal. Appl.*, vol. 4, no. 3, pp. 247–269, 1998.

[192] K. Andra, C. Chakrabarti, and T. Acharya, "A VLSI architecture for lifting-based forward and inverse wavelet transform," *IEEE Trans. Signal Process.*, vol. 50, no. 4, pp. 966–977, 2002.

[193] C.-T. Huang, P.-C. Tseng, and L.-G. Chen, "Flipping structure: An efficient VLSI architecture for lifting-based discrete wavelet transform," *IEEE Trans. Signal Process.*, vol. 52, no. 4, pp. 1080–1089, 2004.

[194] H. Liao, M. Mandal, and B. Cockburn, "Efficient architectures for 1-d and 2-d lifting-based wavelet transforms," *IEEE Trans. Signal Process.*, vol. 52, no. 5, pp. 1315–1326, 2004.

[195] C.-T. Huang, P.-C. Tseng, and L.-G. Chen, "Analysis and VLSI architecture for 1-d and 2-d discrete wavelet transform," *IEEE Trans. Signal Process.*, vol. 53, no. 4, pp. 1575–1586, 2005.

[196] A. Kamboh, M. Raetz, K. Oweiss, and A. Mason, "Area-power efficient VLSI implementation of multichannel DWT for data compression in implantable neuroprosthetics," *IEEE Trans. Biomed. Circ. Syst. I*, vol. 1, no. 2, pp. 128–135, 2007.

[197] D. Donoho, "Unconditional bases are optimal bases for data compression and for statistical estimation," *Appl. Comput. Harmon. Anal.*, vol. 1, no. 1, pp. 100–115, 1993.

[198] M. Vetterli and C. Herley, "Wavelets and filter banks: Theory and design," *IEEE Trans. Acoust. Speech*, vol. 40, no. 9, pp. 2207–2232, 1992.

[199] G. Beylkin, "On the representation of operators in bases of compactly supported wavelets," *SIAM J. Numer. Anal.*, vol. 6, no. 6, pp. 1716–1740, 1992.

[200] M. Holschneider, R. Kronland-Martinet, J. Morlet, and P. Tchamitchian, "A real-time algorithm for signal analysis with the help of the wavelet transform," in *Wavelets, Time-Frequency Methods and Phase Space*. Berlin: Springer-Verlag, 1989, pp. 289–297.

[201] R. Coifman and M. Wickerhauser, "Entropy-based algorithms for best basis selection," *IEEE Trans. Inform. Theory*, vol. 38, no. 2, pp. 713–718, 1992.

Publications Involving The Author

- K. Imfeld, S. Neukom, A. Maccione, Y. Bornat, S. Martinoia, P.-A. Farine, M. Koudelka-Hep, and L. Berdondini, "Large-Scale, High-Resolution Data Acquisition System for Extracellular Recording of Electrophysiological Activity," *IEEE Trans. Biomed. Eng.*, Vol. 55, No. 8, pp. 2064-2073, 2008.

- K. Imfeld, A. Maccione, M. Gandolfo, S. Martinoia, P.-A. Farine, M. Koudelka-Hep, and L. Berdondini, "Real-Time Signal Processing for High-Density MEA Systems," *Int. J. Adapt. Control*, doi:10.1002/acs.1077.

- K. Imfeld, L. Berdondini, A. Maccione, S. Martinoia, P.-A. Farine, and M. Koudelka-Hep, "Hardware-Based Real-Time Signal Processing for High-Density MEA Systems," *Proc. 6th Intl MEA Meeting*, Reutlingen, Germany, July 8-11, 2008.

- L. Berdondini, K. Imfeld, M. Gandolfo, S. Neukom, M. Tedesco, A. Maccione, S. Martinoia, and M. Koudelka-Hep, "APS-MEA Platform for High Spatial and Temporal Resolution Recordings of In-Vitro Neuronal Networks Activity," *Proc. 6th Intl MEA Meeting*, Reutlingen, Germany, July 8-11, 2008.

- A. Maccione, L. Berdondini, K. Imfeld, M. Koudelka-Hep, S. Martinoia, "'Exploring Neuronal Circuitries: a 4096 High-Resolution Microelectrode Array Platform," *Bioengineering 2008*, Imperial College, London, UK, September 18-19, 2008.

- A. Maccione, M. Gandolfo, M. Mulas, L. Berdondini, K. Imfeld, M. Koudelka-Hep, and S. Martinoia, "A New Software Analysis Tool for Managing High Density MEA Systems," *Proc. 6th Intl MEA Meeting*, Reutlingen, Germany, July 8-11, 2008.

- K. Imfeld, A. Garenne, S. Martinoia, M. Koudelka-Hep, and L. Berdondini, "Motivations and APS-based Solution for High-Resolution Extracellular Recording from In-Vitro Neuronal Networks," *Proc. 3rd Intl IEEE EMBS Conf. Neural Eng.*, Kohala Coast, Hawaii, May 2-5, 2007.

- K. Imfeld, A. Garenne, S. Neukom, A. Maccione, S. Martinoia, M. Koudelka-Hep, and L. Berdondini, "High-Resolution MEA Platform for In-Vitro Electrogenic Cell Networks Imaging," *Proc. 29th Annual Intl Conf. IEEE Eng. in Medicine and Biology Society*, Lyon, France, August 23-26, 2007.

- L. Berdondini, P. Massobrio, M. Chiappalone, M. Tedesco, K. Imfeld, A. Maccione, M. Gandolfo, M. Koudelka-Hep, and S. Martinoia, "Extracellular Recordings from High Density Microelectrode Arrays Coupled to Dissociated Cortical Cultures," *J. Neurosci. Meth.*, Vol. 177, pp. 386396, 2009.

- L. Berdondini, M. Chiappalone, P. D. van der Wal, K. Imfeld, N. F. de Rooij, M. Koudelka-Hep, M. Tedesco, S. Martinoia, J. van Pelt, G. Le Masson, A. Garenne, "A Microelectrode Array (MEA) Integrated with Clustering Structures for Investigating In-Vitro Neurodynamics in Confined Interconnected Sub-Populations of Neurons," *Sensors and Actuators B*, vol. 114, pp. 530-541, 2006.

- P. Massobrio, S. Martinoia, L. Berdondini, K. Imfeld, M. Koudelka-Hep, A. Garenne, G. Le Masson, "Recording Electrophysiological Activity by Means of High Density MEAs: Theoretical Models, Extracellular Signal Simulations and Measurements," *Proc. 5th Intl MEA Meeting*, Reutlingen, Germany, July 4-7, 2006.

- S. Generelli, L. Berdondini, K. Imfeld, Y. Bornat, P. van der Wal, O. Guenat, M. Koudelka-Hep, "Microarray-Based Neurointerfaces", presented at *IV World Congress on Biomimetics, Artifical Muscles & Nano-Bio 2007*, Torre Pacheco, Spain, Nov 6-9, 2007.

- L. Berdondini, O. Frey, S. Generelli, O. Guenat, K. Imfeld, P. van der Wal, M. Koudelka-Hep, "Microelectrode Arrays: Technology and Applications," presented at *Swiss Chemical Society, Division of Analytical Chemistry*, E. Pretsch Symposium, Zürich, June 28-29, 2007.

- K. Imfeld, P.-A. Farine, M. Koudelka-Hep, and L. Berdondini, "High-Density Micro Electrode Array for In-Vitro Electrophysiological Activity Monitoring," *Proc. Ann. Meeting of the Swiss Society of Biomedical Engineering SSBS*, Neuchâtel, Switzerland, September 13-14, 2007.

Acknowledgments

This PhD thesis was supported by the IDEA project, Grant FP6-516432.

Many people have supported me during my PhD thesis. All my gratitude goes to:

- Milena, for her extraordinary and continuous support during the IDEA project, as a coordinator of the IDEA project and who allowed me to carry out my PhD thesis on a fascinating topic

- Luca, as a close friend, co-worker on the IDEA project and all the great past and future discussions we had and will have together

- Pierre-André Farine, head of the ESPLAB, who gave me the chance and freedom to work independently in his laboratory

- André Garenne and Steve Tanner as members of the jury

- Alessandro Maccione, Sergio Martinoia, Mauro Gandolfo, Mariateresa Tedesco and all the members of DIBE working on the IDEA project

- Simon Neukom from CSEM, Zürich, and Fanny Farrugia from INSERM, Bordeaux, for the great collaboration during the IDEA pro-ject

- Christian Robert, my "roommate", "Wikipedia" and friend at IMT, for all his technical assistance, especially on IT topics, and all the interesting discussion we had during this time

- Yannick Bornat for his kind support during the time at IMT

- All my collaborators at IMT, especially Marcel Frei, Luca Rossi, Grégoire Wälchli, Alexis Boegli, Guilherme Bontorin and, in particular, Patrick Stadelmann who also assisted me for all questions related to Latex as well as administrative stuff

- Sandrine Piffaretti for setting up very efficiently all the administrative tasks of this PhD thesis

- Joëlle Banjac for her great work and daily support

- Yvan Ferri, Philippe Vez and Arnaud Casagrande for relaxing chats during our "Wednesday Clubs"

- Frédéric Mariotti and Salman Ozserik for their martial arts passion sharing with me during all these years

- My parents who gave me the opportunity to do all this

- **Jutta**, my girlfriend and my heart, for all the love and support every day

Die VDM Verlagsservicegesellschaft sucht für wissenschaftliche Verlage abgeschlossene und herausragende

Dissertationen, Habilitationen, Diplomarbeiten, Master Theses, Magisterarbeiten usw.

für die kostenlose Publikation als Fachbuch.

Sie verfügen über eine Arbeit, die hohen inhaltlichen und formalen Ansprüchen genügt, und haben Interesse an einer honorarvergüteten Publikation?

Dann senden Sie bitte erste Informationen über sich und Ihre Arbeit per Email an *info@vdm-vsg.de*.

Sie erhalten kurzfristig unser Feedback!

VDM Verlagsservicegesellschaft mbH
Dudweiler Landstr. 99
D - 66123 Saarbrücken
www.vdm-vsg.de

Telefon +49 681 3720 174
Fax   +49 681 3720 1749

Die VDM Verlagsservicegesellschaft mbH vertritt

Printed by Books on Demand GmbH, Norderstedt / Germany